A "PICTORIAL" JOURNEY THROUGH THE PAGES OF EVOLUTION

LOU GIAFFO

ISBN: Softcover 978-1-4500-1220-1

This book was printed in the United States of America.

To order additional copies of this book, contact:
Xlibris Corporation
1-888-795-4274
www.Xlibris.com
Orders@Xlibris.com

In Memory of

SID SHEMONSKY

1930-2008

CONTENTS

LIST OF ILLUSTRATIONS

FOREWARD

Without Life the Universe has No Meaning!

The Theory of Evolution is still very much a controversial issue with many people, with at least 50% of Americans, who still don't believe in the theory, mainly on religious grounds, despite its living proof now just about proven through the science of DNA genetics.

TV Shows like *Forensic Files* and *Coldcase Files* have, more than anything else, heightened awareness of this in documenting the role DNA testing has played in solving heretofore unsolvable cases which had gone cold for decades and more, in finding those falsely accused of crimes, innocent, along with convicting those who are found guilty.

My only purpose, or I should say – my "mission," for writing this book on evolution is to throw some *light* on the subject, obscured through the *heat* of controversy between evolution and religion, through laying out the facts sequentually, starting with the beginning of life some 3 ½ billion years ago. I do this more in the role of a teacher (which I was and still am) than a writer.

When Charles Darwin first introduced the concept of evolution in his *The Origin of the Species,* contending that through *Common Descent*, all species, including Man, had descended from lower orders of ancestor, it sparked a bitter controversy that goes on to this day.

The Catholic church, the first major religious order to do so, has come to terms with this and has reconciled itself with this issue, rationalizing it as the workings of the divine rather than by the random chance of nature through evolution. The Intelligent Designers (Creationists) accept the theory of evolution, more or less as it stands, but with the proviso that some of the things in nature are too detailed and complicated to be left to the chance of evolution, and could only be the work of an Intelligent Designer: God!

But as Shakespeare said, "Whats in a Name?" Any architect, be it the evolutionary architect or a Devine architect, has to start with a plan, or blueprint, if you will, to build a structure, as shown by the structure of our own bodies, through a blueprint called DNA. The story of life is the same only drawn on the board of a much larger scale, a scale of billions of years.

Meaningfull life began more than three and a half billion years ago, starting with simple single-celled *prokaryote* bacteria, which then evolved into multi-celled *eukaryite* bacteria two billion years later, which then evolved into the varied animal and plant life forms of today.

Our vertebrate ancestors started as lobe-finned fishs, then evolved into amphibians, then into tetrapods (four-legged reptiles), then dinosaurs, then mammals (our forebares: parsimians, monkeys, apes, and us, Man), and birds (living descendents of dinosaurs), which are all graphically layed out in this work.

Some of us will look with contempt at our humble forebare origins. But the inescapable fact is, that without their resourcefulness, we wouldn't be here to tell this story of life. In their noble struggle for survival, they prevailed against the impossible odds, those of a hostile environment, saturated with a myriad of fierce predators. Their survival is their everlasting legacy to us.

ABOUT THE AUTHOR

A second generation Albanian/American, Lou was first of four children, born in Boston, Massachusetts to Kasem and Ilfan Giaffo, but was raised in Hartford, Connecticut (about 100 miles southwest of Boston), shortly relocating there after Kasem sold out his share of interest in a restaurant partnership with three other Albanian Americans.

It wasn't too long before Kasem opened up his own restaurant in midtown Hartford, called the "Midway Lunch." It became a family business in all sense of the word, and a home-away-from-home. While Kasim did all of the cooking ("the nearest thing to home cooking," he claimed, in a sign prominently displayed in bold letters), Ilfan served both as a waitress and cashier, Lou and brother Chufi (two years his junior) served on KP duty, peeling spuds (potatoes), in rat the invested cellar since knee-high. A relative of the family's, Vesel, served as the short-order cook. The only outsiders were the dishwashers, with a high turnover rate.

Although Kasem's cuisine was rather limited, what he did offer was vary good, particularly his daily specials; as a result, he built up a loyal and steady clientele. Following the passing of grandma, Aba, the matriarch of the family on the father's side, who arrived in America in her late 60's (although she never learned to speak English, it didn't stop her from communicating with others, but served, on the other hand, compel the young Giaffo's to learn Albanian in order to communicate with her), the Giaffo's decided to make a trip to Albania, where she was also beloved, to share their grief with the multitude of their relatives. despite the fact that Albania was, at that time, in the late 1950s, was under the iron grip of Communist dictator, Enver Hoxhia Since the Albanian Communist regime had no diplomatic relations with the United States, the doors were closed to American tourists, including even Albania/American tourists. But through the influence of a cousin they had high up in the communist regime, however, the Giaffo's were able to obtain a tourist visa.

In late 1959, the Giaffo family (Lou, Chufi, Lule, Vera, Ilfan & Kasem), boarded the Italian ocean liner, Christoforo Columbo (Christopher Columbus), in New York harbor. Following 5 days of fun and frolic, on the high seas, the liner docked at the port of Naples, Italy. From there Kasem commissioned a taxi to take the family to Rome (Roma) about 100 miles southeast.

Squeezed into the small cab like sardines, the Giaffo's headed for Rome, on the ancient Appian Way road way, taking in the intoxicating scenery along the way, and making occasional pit stops to eat and refresh. These pit stops also offered the Giaffo's the opportunity to get acquainted with the local folk and take pictures, mainly by Lou, an incurable camera bug, who used both a still camera and a super-8 movie camera.

Since the Giaffo's had to wait a couple of weeks for the boat to Albania to depart from the Italian port of Bari across the Adriatic Sea to the Albanian port of Durres. They made good use of the interim for sightseeing of the historic sites, in and around Rome, and of course capturing it all on film.

Because of the sickening prevailing pungent smell of gasoline, the overnight trip on the Adriatic Sea from the port of Bari, Italy to the port of Durres, was a sleepless one for the Giaffo family. The dock workers couldn't stop gawking at the tired, bleary eyed Giaffo's (as if they had just dropped from another planet) as the boat slowly sailed past them. This of course is not hard to understand since they (Giaffo's) were the only outsiders to visit Albania in over 15 years of Communist rule.

Most of the rest of the day was spent in the Albanian customs (Dugon) office, as the Giaffo's had to endure close interrogation scrutiny, along with standing by helplessly as the customs officials opened and unceremoniously dumped the contents of their luggage all over the floor, after which the Giaffo's had to pick up and repack the luggage again.

When the inspection process was finally completed at day's end, customs allowed the Giaffo's to leave and greet their multitude of relatives who had been gathering outside of the custom's building throughout the day.

The crowd of relatives immediately mobbed the Giaffo's and literally lifted off their feet and dumped them into two waiting ancient pickup trucks. Packed tightly between relatives, the Giaffo's were barraged with various questions, which were incomprehensible since they (relatives) were all talking, without a pause, at the same time.

As the ancient trucks rumbled along the roadway in the ink black moonless night towards Tirana, the capital city of Albania about 30 miles due north, the seemingly spring less trucks didn't miss a pothole, leaving the already weakened stomachs of the Giaffo's from the boat trip in state of regurgitation, which was expressed before they reached their destination.

When the depleted Giaffo's reached Tirana, they were greeted by even more relatives, all of whom, with typical Albanian hospitality, invited them again shouting in unison, to their homes with food and drink already lay in waiting on the tables. Although their stomach's were empty from regurgitations, food was the furthest thing from the Giaffo's mind at that time.
Needless to say, the Giaffo's were in a quandary as to which invitation to accept, but before they could come to a decision, they found themselves involuntarily set in an anonymous living room. One good thing – it spared them the necessity of a decision.

The festivities continued round the night, in the myriad of their relative's homes, in an orgy of nonstop eating and drinking (mainly by the relatives). This, however, was interrupted in one of the homes got so overloaded with wall-to-wall revelers that the floor gave way under all the weight, plunging them through the splintered floorboard in a magic-like disappearing act. Luckily, except for a few minor bruises and cuts, there were no major injuries!

In the next two months, the Giaffo's made it a point to visit with all of their relatives, both on the father's and mother's side, along with taking in all of the most important Albanian historical sites, the highlight of which was the mountaintop fortress Kruje, near Tirana, where the Albanian National hero, Skanderbeg, fought off and defeated the Ottoman Turkish hordes under Mohammed 2nd in every battle, usually against overwhelming odds of 15 to 1. In the end, the Turks were so impressed with Skanderbeg's military genius, that they attributed it more to a sort of magic. So ten years after his death in 1468, the Turks exhumed his body in 1478. They then cut off several of his body parts to save as amulets (magic charms) in the hopes that some of his magic military genius would rub off on them.

For the trip back home, the Giaffo's boarded the American ocean liner "Constitution" home, stopping off at various 'ports of call' in the Mediterranean Sea, including Casablanca, Morocco, Barcelona, Spain, and the Island of Madura, off the coast of Spain, etc. Except for one storm when nearing New York City, it was smooth sailing. The timing of the storm, which couldn't have been worse since it hit on New Year's Eve, which was celebrated twice because of crossing double time zones. With the exception of a few hardy stomachs, including Lou's and his father, Kasem, and a few other passengers, who were able to make for breakfast, the ship's dining hall was made up of mostly of empty chairs the next morning.

Kasem opened up another restaurant in Hartford, Connecticut called the "Deity," named after the Deity mountains, which surround the capital city of Tirana. But the restaurant went out of business a few years later. This gave Lou a chance to strike out on his own in pursuit of his long desired art career.

Ten years earlier, although an 8th grade dropout, Lou was nevertheless accepted in the U.S. Marine Corps, and served a three stint. He was successful artist for over 16 years in New York City ("The Big Apple").

His early dropout from school was due mainly to health issues, such as weak eye muscles where he couldn't focus his eyes on the subject for more than a few minutes at a time before his vision blurred. So he had to go through a continuous cycle of adjustment and readjustment with a few seconds of rest in-between. This of course, didn't exactly make for speed-reading. Nobody in the school system, however, thought of giving him an eye examination. On top of this He was also laid up for a whole year with Plurisy (fluid in the lungs).

Lou took some correspondence courses through the Marine Corps Institute while serving with the U.S. Marine Corps at Camp Lejeune, North Carolina, through the urging of one of his superior officers. Because of his early negative educational experiences, Lou didn't place much stock in his educational prowess. But to his surprise, however, his first surprise, his first grades came back with straight A's. That was the first big step in restoring his confidence for future pursuits in education.

Although working successfully as a commercial artist for 16 years, Lou still felt incomplete so he decided to formalize his educational status, and took and passed his high school equivalency test for his diploma. He then immediately enrolled in Queensborough Community College, New York, for his associate's degree, then going across town to Queens College, in Queens New York, for his bachelor's degree. Lou earned masters degree in education at Austin Peay State University, in 1980, located in Clarksville, Tennessee, where he still lives. Lou earned all of his degrees within a ten-year period on a part-time bases, while working full-time.

After earning his master's degree in education Lou changed careers and went into teaching. Although this book, A "Pictorial" Journey Through The Pages Of Evolution, is Lou's 7th published book, Lou considers himself more as a teacher than a writer, more or less using as another teaching tool.

Lou lost his long bachelorhood on a trip to Albania in 1991, to visit his relatives, following the fall of Communism. They (relatives) introduced him to Meliha Lofca, where they shortly became engaged. Lou's 3-week visa an out and he had to return home again. But he went back to Albania to pick up his future wife about six months later.

Jason Lou was born on June 18, 1993, and is now a 15 year old high school student, excelling both in the academic area as a straight-A student, and in the music, where he has mastered the saxophone, clarinet, base guitar, and electric guitar. He also sings and writes music.

INTRODUCTION

Life originated more than **3. 5 billion years** *ago* in the sea. It stayed there for the majority of the history of Earth.

The common ancestor of all life gave rise to *prokaryotes* (“ordinary” bacteria). Bacteria are the only life forms found in the rocks for a long, long, time!

Animals start appearing prior to the Cambrian, about **600 million years ago**. Animal life evolved and flourished only after plants became available as food. Like animals today, prehistoric animals depended on green plants. Green plants can use energy from sunlight to make food. This is called *photosynthesis*, and this **evolved** around **3.4 billion years ago**. Animals cannot get energy from the sunlight to make food, so they must eat either plants or plant-eating animals.

The first animals dating from just before the Cambrian were called the *Ediacarian fauna*. It is unclear if these forms have any surviving descendents. Some look a bit like *Cnidarians* (jellyfish, sea anemones and the like) – others resemble annelids (earthworms).

All the *phyla* (the second highest taxonomic category) of animals appeared around the Cambrian. The animals of the *Burgess shale* are the example of Cambrian animal **fossils**. These fossils from Canada, show a bizarre array of creatures, some of which appear to have unique body plans unlike those seen in living animals.

All these creatures were *invertebrates* – that is, they had *no backbones*. Their soft bodies also lacked shells or other body parts. As a result, few of them were preserved as fossils.

According to fossils found by scientists, animals first became abundant at the beginning of the *Paleozoic Era*, about **570 million years ago**. The first skeletons developed in a number of animal lineages, using the hard shells or tough outer frames to protect them from enemies.

Trilobites ranked among the most common early Paleozoic animals. They were ancient *arthropods*, the group to which spiders and crabs belong. Most trilobites fed on small food particles from the sand or mud on which they lived. Such food may have Included *small worms* and other bottom-dwelling animals. Another common group of Paleozoic animals, the *brachiopods*, had shells similar to those of *claims*. They lived on the sea floor or burrowed in the mud. Brachiopods ate tiny organisms. They fed by opening their shells and filtering food from the water with a comb-like organ.

The first animals with backbones, called *vertebrates*, probably appeared at the end of the Cambrian Period, more than **500 million years ago**. The earliest vertebrates were *small*. Plates of bone usually protected their heads and much of their bodies. In later vertebrates, a skeleton made up of many bones formed inside the body. The skeleton provided a solid internal frame to which muscles could attach.

The earliest vertebrates had mouths without jaws. They most likely fed on small bits of dead animals or creatures on the sea floor or in the water. In the *Silurian Period*, beginning **440 million years ago**, vertebrates developed *bony jaws* and *teeth*. These mouth parts enabled them to catch and consume larger kinds of food. The vertebrates also developed movable *paired fins*, enabling them to become more active swimmers.

The Cambrian fauna (*trilobites, inarticulate brachiopods*, etc.) declined slowly during this time.

The *Devonian Period*, often called the *Age of Fishes*, began about **410 million years ago**. One of the largest Devonian fishes, *Dunkleostaus* grew 23 feet (7 meters) long. Massive plates of bone protected its head and the front of its body. Its jawbones formed *sharp ridges* for cutting up and eating fish.

Bony fishes of the Devonian Period had to types of fins, *rayed* and *lobed*. Rayed fins consist of a web of skin supported by a skeleton of rods called *rays*. Fish with this type of fin, known as *ray-finned fishes*, are fast swimmers. Lobed fins consist of a fleshy stalk fringed with rays. The Devonian Period was the time of greatest abundance for fishes with this type of fin, called *lobe-finned fishes*. Lobed fins enabled the creatures to crawl along the bottom of the sea or over land. Many lobed-finned fishes also had *lungs* with which they gulped air when there was *not enough oxygen* in the water. All land-living vertebrates, including you and I (*human beings*), descended from these fishes.

Plants **evolved** from ancient *green algae* over **400 million years ago**. *Plants* and *fungi* (in symbiosis) invaded the land roughly at the same time. The first plants were *moss-like* and required moist environments to survive. Later, evolutionary developments such as a *waxy cuticle* allowed plants to exploit more inland environments. Still mosses lacked true *vascular tissue* to transport *fluids* and *nutrients*. This limited their size since these must diffuse through the plant. Vascular plants evolved from *mosses*. The first vascular land plant known was *Cooksonia*, a spiky, branching, leafless structure.

At the same time, or shortly thereafter, *arthropods* followed plants onto the land. The first land animals were known as *myriapods, centipedes, scorpions, spiders, mites*, and *millipedes*. Living scorpions belong to an ancient group which dates back about **400 million years**, alongside *sharks*.

The earliest known *coelacanth* appeared **350 million years ago**. These fish were thought to be extinct, but recently many living coelacanths have been discovered. *Lungfish* have fossil relatives which date back also to this time (**350 million years ago**).

The Sturdy *fins* of lobe-finned fishes evolved into the *legs* of land-living vertebrates, with *ankles, wrists, fingers*, and *toes* suitable for leaving the water. *Lungs* became a fulltime *oxygen supply*.

Four-footed land vertebrates are known as *tetrapods*. The first tetrapods, called *amphibians*, had to return to the water to lay eggs. These eggs were enclosed in a jelly—like substance that, if left on land, would dry out and *kill* the eggs. Thus amphibian eggs developed and hatched in water, where the young lived until they eventually adults, and evolved into *land-dwelling*, and *air-breathing species*.

Ichthyostega, an amphibian, is the first among the first-known land vertebrates. It was found in Greenland and was derived from lobe-finned fishes called **Rhipidistians**.

During the *Late Paleozoic* Era, a great variety of amphibians inhabited the shores of lakes and rivers, dividing their lives between land and water. Modern amphibians, including *frogs* and *toads*, are small, but some Paleozoic amphibians were large. The meat-eating **Eryops** grew more than 5 feet (1.5 meters) long and had a *big head*. Others, such as *Ophiderpeton*, were *small* and *snakelike*.

The earliest **insects** were *wingless* and appeared during the *Devonian Period*. By the late *Carboniferous Period*, about **300 million years ago**, many insects had developed *wings*. They included *Meganeure*, a spectacular animal resembling a dragonfly, which had a wingspan of up to 26 inches (*65* centimeters).

Amphibians gave rise to *reptiles* during the *Carboniferous Period*. Reptiles had evolved *scales* to decrease water loss and a *shelled* egg, permitting the young to be hatched on land. Special membrane inside the eggs enclosed and protected the developing young in a fluid-filled chamber. This type of egg enabled reptiles to live entirely on land. Among the earliest well-preserved reptiles is **Hyonomus**, discovered from rocks in Nova Scotia.

Therapsids derived from *Pelyosaurs* appeared **286 million years ago** in the *Permian*.

Some were *herbivores* (plant-eaters) and some were *carnivores* (meat-eaters). The herbivorous *therapsids*, or *anomodonhs*, were variable in size, form, and dentition (teeth). Important in their time, they represented an evolutionary *blind alley*.

At the end of the *Paleozoic Era*, the earth's climate generally grew warmer and drier. *Deserts* spread over large areas, displacing many amphibians that needed water for their survival. At the end of the period, about **248 million years ago**, the seas shrunk as sea levels dropped severely worldwide. This lowering of the sea level helped cause one of the *largest known extinctions*. *Ninety-five percent* of species may have perished in the seas at that time.

The last of the *Cambrian Fauna* went extinct. The *Paleozoic fauna* took a nose dive from about 300 families to about 50. It is estimated that 90% of all species (50% of all families) met their end. Following this event, the Modern fauna, which had been slowly expanding since the *Ordovician*, took over. The modern faunas included *fish, bivalves, gastropods* and *crabs*. These were barely affected by the Permian extinction. The modern fauna subsequently increased to over *600 marine families*, up to the present time. The Paleozoic fauna held already at about *100 families*. A *second extinction* event shortly following the *Permian* kept animal diversity low for awhile.

After the Permian extinction, *gymnosperms* (ex-pines) became more abundant. Gymnosperms had evolved *seeds*, from seedless fern ancestors, which helped their ability to *dispense*. Gymnosperms also evolved *pollen*, enclosed sperm which allowed for more *out-crossing*.

The earth's greatest extinction event ever, the Permian extinction **248 million years ago**, decimated *animal life*, but left the land ripe for domination of certain animals. Reptiles rose to the challenge, evolving into a wide variety of species including **archosaurs**, the group that spawned *dinosaurs, pterosaurs*, and *crocodilians*.

Only *4* of the *23* orders of *reptiles* that have *ever existed* have living representatives. The best-known of the 23 orders include **Squamata,** the largest order of living reptiles, containing more than 5000 species of *lizards* and *snakes*; *Chelonia*, or *Testudines* the *turtles* and *tortoises;* **Crocodiles** represent the closest living relatives of *dinosaurs, birds*, and *Rhynchocephalia*, all species now extinct, except for the New Zealand *tuatara*.

During the *Triassic Period*, all the continents of the earth were joined together to form a *super-continent* called **Pangaea**. The continent was so large, that much of the interior was a long, long, walk to the sea, which resulted in *very hot* and *arid* climates.

The *carnivorous* **therpasids** gave origin to the *mammals* in the *Triassic*, **248 million years ago**. The carnivorous *therapsida*, or *thenriodontia*, developed *mammalian* characteristics. The complete *transition* to mammals depended upon methods of *reproduction*, the ability to nurse the young, development of *hair* and *warm-bloodedness*, and other features of which skeletal remains, not surprisingly, left no *fossil* evidence.

230 million years ago, the plant-eating **ductodon** (a mammal-like reptile) lived. Their skulls were *squat* and *pig-like*.

Procolophon, a small early reptile, fed on *tubers*.

In the *Early Triassic*, however, *ecosystems* were dominated by the immediate ancestors of the dinosaurs, the **archosaurs**. Some archosaurs developed a *bipedal* (two—legged) posture and *S-shaped* necks, and it was some species of these that eventually evolved into dinosaurs. These earliest dinosaurs were about *0.5 meters* (about 1.6 ft.) long. By about **175 million years ago**, most of the basic varieties of dinosaurs had made their appearance.

Those *archosaurs* most closely related to the *dinosaurs* were forms such as **Marasuchus**. By the Late Triassic, several early *theropods* are known, as the dinosaurs rapidly diversified. Those dinosaurs, such as *Eoraptor* (**225 million years ago**), *Coelopysis* (**231-213 million years ago**) and *Herrereasaurus* were all *carnivors* and, despite their diversity, were quite *rare* at this time:

* An *Eoraptor* skull and skeleton, about 1 m (about 3 feet) long discovered in northwestern Argentina in 1992 indicates that *Eoraptor* had sharp teeth, two legs, three-fingered hands, and three-toed feet.

* *Coelophysis* was a small, meat-eating, early dinosaur. *Coelophysis* was slender and fast, with long jaws filled with dagger-like teeth. About 3 m (10 feet) long, it inhabited highland forest areas. Many complete *Coelophysis* skeletons have been found in New Mexico, and several thousand dinosaur fossils may still remain there. Scientists believe that these dinosaurs probably died in a *flash flood* or in a *mud-trap*.

210 million years ago, *Riojasuchus* was one of the first *thecodonts* or "socket-toothed" reptiles. It was built like long-legged crocodile, and had powerful teeth and jaws.

5 million years ago, the last *mammal-like reptiles* that appeared just before the early dinosaurs, was *Massetognathus*, large and doglike in appearance.

200 million years ago:

* The furry Megazostodon lived. It was one the earliest true mammals.

* Ichthyosaus lived at this time as well. They were swimming reptiles, having long, narrow, pointed snouts.

* The omnivourous *Thecodontosaurus* also lived at this time.

  Dinosaur's closest living relatives are *crocodiles*, which date back to this time, too.

  One modification that may have been a key to their success was the evolution of an *upright stance*. *Amphibians* and *reptiles* had a *splayed stance* and walked with an *undulating pattern* because their limbs were modified from fins. Their gait was modified from the *swimming movement* of fish. Splay stance animals could not sustain continued locomotion because they couldn't breath while they moved, their undulating movement compressed their chest cavity. Thus, they had to stop every few steps to breath before continuing on their way.

155 million years ago – *plesiosaurs* lived, a fierce marine reptile.

150 million years ago – the first *birds* appeared.

147 million years ago – the *sphenodontid lizard* lived.

145 million years ago – *pterodactylus* flew the skies. *Somepterosaurs* the size of sparrows; others were as big as small aircraft.

149 million years ago – king crabs (only remotely related to modern crabs) and *gryolus* (bony fish) lived.

115 million years ago – the plant-eating *iguanodons* thrived.

110 million years ago – *gastropods* (snails) lived.

100 million years ago – some marine groups, like lobsters, were barely affected by mass extinction.

95 million years ago – *turtles* were another group that survived mass extinction.

90 million years – the *Albertosaurus* lived alongside bony ray-finned fish.

85 million years ago –*marsuppials* (kangaroo-like) lived.

75 million years ago – crabs, like lobsters, survived the extinction.

70 million years ago – *mosasaurs* (giant marine lizards) used its large-pointed teeth to crack open shells of animals and ammonites.

65.4 million years ago – the end of the Cretaceous is marked by a *minor mass extinction*. This extinction marked the demise of all the *lineages of dinosaurs*, save the birds.

55 million years ago – up to this point *mammals (tree Shrews, prosimians, tarsiers, etc.)* were confined to *nocturnal, insectivorous* niches. Once the dinosaurs were out of the picture, they diversified and flourished.

50 million years ago – horses appeared

35 million years ago – gnawing rat-like rodents appeared

34 million years ago – *monkeys e*volved from *prosimian – gnawing rodent-like shrews.*

21 million years ago – *apes* evolved from monkeys.

6 million years ago – hominids evolved from apes.

200,000 years ago – *homo sapiens* appeared

EOLOGICAL TIME

Era	Millions of years ago
Precambrian Time	
Archean Era	
Proterozoic Era	2500-5704
Phanerzoic Time	
Paleozoic Era	
Cambrian Period	570-505
Ordovician Period	505-430
Silurian Period	438-408
Devonian Period	408-360
Carboniferous Period	360-286
Permian Period	286-245
Mesozoic Era	
Triassic Period	245-208
Jurassic Period	208-144
Cretaceous Period	122-66.4
Cenozoic Era	
Tertiary Period	66.4-0-2
Paleogene	58-66
Eocene	07-58
Oligocene	24-37
Miocene	05-24
Pliocene	02-5
Pleistocene/Holocene	00-2

INTRODUCING EVOLUTION

In biology, evolution is the process of change in the inherited traits of a population of organisms from *one* generation to the *next*. The *genes* that are passed on to an organism's *offspring* produce the *inherited traits* that are the *basis* of evolution.
Mutations in genes can produce *new* or *altered* traits in individuals, resulting in the appearance of heritable *differences* between *organisms*, but new traits also come from the t*ransfer* of genes between populations, in migration, or between species, in *horizontal* gene transfer. In species that reproduce *sexually*, new combinations of genes are produced by genetic *recombination*, which can *increase* the *variation* in traits between organisms. *Evolution* occurs when these heritable differences become more *common* or r*are* in a population.

There are *two* major *mechanisms* driving evolution: The *first* is *natural selection*, a process causing heritable traits that are *helpful* for *survival* and *reproduction* to become more common in a population, and *harmful* traits to become more *rare*. This occurs because individuals with *advantageous* traits are more *likely* to reproduce, so that *more* individuals in the next generation inherit these traits. Over many generations, *adaptations* occur through a *combination* of successive, small, random *change* in traits, and in *natural selection* of these variants *best-suited* for their *environment*. In an independent process, genetic drift produces *random changes* in the *frequency* of traits in a population. The s*econd* is *Genetic drift*, which results from the role *probability* plays in whether a given trait will be passed on as individuals survive and reproduce. Though the changes produced in any generation by *drift* and *selection* are *small*, differences accumulate with e*ach subsequent* generation and can, *over time*, cause substantial *changes* in the organism.

One *definition* of a species is a group of organisms that can reproduce with one another and produce *fertile* offspring. When a species is separated into populations that are prevented from interbreeding, *mutations, genetic drift*, and *natural selection* cause the a*ccumulation* of differences over generations and the emergence of *new species*. The s*imilarities* between organisms suggest that all known species are descended from a c*ommon ancestor* (or ancestral gene pool) through this process of *gradual divergence*.

Evolutionary biology documents the fact that evolution occurs, and also develops and tests theories that explain why it occurs. Studies of the fossil and diversity of living organisms had convinced most scientists by the mid-nineteenth century that species c*hanged* over time. The mechanism driving these changes, however, remained unclear until the 1859 publication Charles Darwin's *On the Origin of Species,* detailing the *theory* of evolution by *natural selection*. Darwin's work soon led to overwhelming acceptance of evolution *within* the scientific community. In the 1930s, Darwinian *natural selection* combined with *Mendelian inheritance* to form the modern *evolutionary synthesis*, in which the connection between the *units* of evolution (*genes*) and the *mechanism* of evolution (*natural selection*) was made. This powerful explanation and predictive theory directs research by constantly raising new questions, and has become the central organizing principle of modern biology, providing a unifying explanation for the d*iversity* of life on Earth

Heredity

Evolution in organisms occurs through changes in *discrete* traits – particularly characteristics of an organism. In humans, for example, *eye color* is an *inherited* characteristic, which individuals can inherit from *one* of their parents. Inherited traits are *controlled* by genes and the complete set of genes within an organism's *genome* is called its *genotype.*

The complete set of observable traits that make up the structure and behavior of an organism is called its *phenotype*. These traits come from its genotype with the environment. As a result, not *every* aspect of the organism's phenotype is inherited.
Suntanned skin results from the interaction between a person's *genotype* and *sunlight*; thus, suntans are *not* passed on to people's children. People, however, have different r*esponses* to sunlight, arising from *differences* in their genotype; a striking example is individuals with the inherited trait of *albinism,* who do *not* tan and are highly *sensitive* to sunburn.

Heritable traits are propagated between generations via DNA, a molecule, which is capable of *encoding* genetic information. DNA is a *polymer* composed of four types of bases. The sequence of bases along a *particular* DNA molecule *specify* the genetic i*nformation*, in a manner akin to a sequence of letters specifying a text or a sequence of bits specifying a computer program. *Portions* of a DNA molecule that specify a *single* functional unit are called *genes*; different genes have different structure sequences of bases. Within cells, the long strands of DNA *associate* with *proteins* to form condensed structures called *chromosomes*; the region of the chromosome at which a particular gene is located is called a *locus*. If the DNA sequence at a locus *varies* between individuals, the different forms of this sequence are called *alleles*. DNA sequences can change through *mutations*, producing *new* alleles. If a mutation occurs *within* the gene, the new allele may affect the trait that the gene controls, *altering* the genotype of the organism. While, however, this *simple correspondence* between an allele and a trait works in *some* cases, *most* traits are *more complex* and are *controlled* by multiple interacting genes.

Variation

An individual organism phenotype results from both its *genotype* and the influence from the *environment* it has lived in. A substantial part of the variation in phenotypes in the population is caused by the *differences between* the genotypes. The modern evolutionary synthesis defines evolution as the *change* over *time* in this genetic *variation*. The f*requency* of one particular allele will *fluctuate*, becoming more or less *prevalent* relative to other forms of that gene. Evolutionary forces act by driving these changes in allele frequency in *one* direction or *another*. Variation *disappears* when an allele reaches the point of *fixation* – when it either *disappears* from the population or *replaces* the ancestral allele entirely.

Variation comes from mutation, migration between population (*gene flow)*, and the r*eshuffling* of genes through *sexual* reproduction. Variation also comes from exchanges of genes between different species; through, for example, *horizontal* gene transfer in bacteria, and *hybridization* in plants. Despite the constant introduction of variation through these processes, most of the genome of the species is *identical* in all individuals of *that* species. Even relatively *small* changes in genotype, however, can lead to dramatic changes in phenotype: *chimpanzees* and *humans* differ in only about *5%* of their genomes.

Mutation

Genetic *variation* comes from *random mutations* that occur in the genomes of organisms. Mutations are *changes* in the DNA *sequences* of a cell's genome and are caused by r*adiation, viruses, transposes* and *mutagenic* chemicals, as well as errors that occur during *meiosis* or DNA replication. These *mutagens* produce several different types of change in DNA sequences; these can either have *no effect, alter* the product of a gene, or p*revent* the gene from functioning. Studies in the fly *Drosophila melanogaster* suggest that if a mutation changes a protein produced by a gene, this will probably be *harmful*, with about *70 percent* of these mutations having damaging effects, and the remainder being either *neutral* or weakly *beneficial*. Due to the *damaging* effects that mutations can have on *cells*, organisms have evolved mechanisms such as DNA *repair* to remove mutations. The optional mutation rate for a species, therefore, is a *trade-off* between costs of a *high* mutation rate, and DNA *repair enzymes*. Some species such as *retroviruses* have such high mutation *rates* that most of their *offspring* will posses a *mutated* gene.

Such *rapid* mutations may have been *selected* so that these viruses can constantly and rapidly *evolve*, and thus *cancel out* the responses of the human *immune system*.

Mutations can involve large sections of DNA becoming *duplicated*, which is a major source of *new* material for evolving *new* genes, with tens to hundreds of genes duplicated in animal genomes every million years. Most genes belong to *larger* families of genes of *shared ancestry*. Novel genes are produced either through *duplication* and *mutation* of an *ancestral* gene, or by *recombining* parts of different genes to form *new* combinations with *new* functions. The human eye, for example, uses *four genes* to make structures that sense *light*: three for *color vision* and one for *night vision*. All four arose from a single *ancestral* gene. An advantage of duplicating a gene (or even an entire genome) is that *overlapping* or *redundant* functions in multiple genes allows alleles to be *retained* that would otherwise be *harmful*, thus increasing genetic *diversity*.

Changes in chromosome number may involve even larger mutations, where long segments of the DNA within chromosomes break and then *rearranges*. Two chromosomes in the *Homo* genus, for example, *fused* to produce human **chromosome 2**.

This fusion did not *occur* in the lineage of the other *apes*, and they retain their *separate* chromosomes. In evolution, the most important role of such chromosomal *rearrangements* may be to *accelerate* the divergence of a population into *new species* by making populations *less likely* to interbreed, and thereby preserving *genetic differences* between these populations.

Sequences of DNA that can move about the genome, such as *transposons*, make up a major portion of the genetic material of plants and animals, and may have been important in the evolution of genomes. More than a million copies of the *Alu* sequences, for example, are present in the human genome, and these sequences have now been recruited to perform such functions as regulating *gene expression*. Another effect of these mobile DNA sequences is that when they move within a genome, they can either *mutate* or *delete* existing genes and thereby produce genetic diversity.

Sex and recombination

In *asexual* organisms, genes are *inherited* together, or *linked*, as they cannot *mix* with genes in *other* organisms during reproduction. The *offspring* of sexual organisms, however, contain random mixtures of their parents' chromosomes that are produced through *independent assortment*. In the related process of genetic recombination, sexual organisms can also exchange DNA between two matching chromosomes. *Recombination* and *re-assortment* do not *alter* frequencies, but instead *change* which alleles are associated with each other, producing offspring with *new* combinations of *alleles*. While this process *increases* the variation in any individual's offspring, genetic mixing can be predicted to either have *no effect, increase*, or *decrease* the genetic variation in the population, depending on how the various alleles in the population are distributed. If, for example, two alleles are *randomly* distributed in the population, then *sex* will have *no effect* on *variations*; if two alleles tend to be found as a *pair*, however, then genetic mixing will *even out* this *non-random* distribution and over time, make the organism in the population more similar to each other. The overall effect of sex on natural variations remains unclear, but recent research suggests that sex usually *increases* genetic variations and may increase the *rate* of evolution.

Recombination allows even alleles that are close together in a strand of DNA to be inherited *independently*. The rate of recombination is *low*, however, since in humans in a stretch of DNA one million base pairs long, there is about a *one in a hundred* chance of a recombination event occurring per generation. Genes, as a result, close together on a chromosome may not always be shuffled away from each other, and genes that are *close* together tend to be *inherited* together. This tendency is measured by finding out how *often* two alleles occur together, which is called their *linkage in-equilibrium.* A set of alleles that is usually inherited in a group is called a *haplotype*.

Sexual reproduction helps to *remove* harmful mutations and *retain* beneficial ones.

Consequently, when alleles cannot be *separated* by recombination – such as in mammalian Y chromosomes, which pass intact from fathers to sons – *harmful* mutations accumulate. In addition, recombination and re-assortment can produce individuals with *new* and *advantageous* gene combinations. These positive effects are balanced off by the fact that this process can cause mutations and separate *beneficial* combinations of genes.

Population genetics

Evolution, from a genetic viewpoint, is a *generation-to-generation change in the frequencies of alleles within a population that shares a common gene pool.* A *population* is a *localized* group of individuals belonging to the *same* species. All of the *moths* of the *same* species living in an isolated forest, for example, represent a *population*. A *single* gene in this population may have several alternate forms, which account for variations between the phenotypes of the organism: An example might be a gene for *coloration* in moths that has two *alleles*, **black** and **white**. A *gene pool* is the *complete* set of alleles in a *single* population, so *each* allele occurs a certain number of times in a gene pool. The fraction of genes within the gene pool that form a particular allele is called the *allele frequency*. Evolution occurs when there are changes in the frequencies of alleles within a population of interbreeding organisms; the allele for *black color* in a population of moths, for example, becomes more common than the *white one*.

To understand the mechanisms that cause a population to evolve, it is useful to consider what conditions are required for a population not to evolve. The *Hardy-Weinberg principle* states that the frequencies of alleles (variations in a gene) in a sufficiently large population will remain constant if the only forces acting on that population are the random *reshuffling* of alleles during the formation of the sperm or egg, and the random *combination* of the alleles in those *sex cells* during fertilization. If such a set in the population is said to be in the *Hardy-Weinberg equilibrium* – it is not evolving.

Mechanisms

There are three basic mechanisms of evolutionary change: (1) *natural selection*: (2) gen*etic drift*, and (3) *gene flow*. Natural selection *favors* genes that *improve* capacity for *survival* and *reproduction*. Genetic drift is *random* change in the frequency off alleles, caused by the *random sampling* of a generation's genes during reproduction, and gene flow is the *transfer* of genes within and between populations. The relative importance of n*atural selection* and *genetic drift* in a population varies depending on the strength of the selection and the effective population size, which is the number of individuals capable of b*reeding. Natural selection* usually predominates in *large* populations, while genetic drift dominates in *small* populations. The dominance of genetic drift in small populations can even lead to the *fixation* of slightly deleterious *mutations*. As a result, *changing* population size can dramatically influence the course of evolution. *Population* bottl*enecks*, where the population *shrinks* temporarily and therefore *loses* genetic variation, result in a more *uniform* population. Bottlenecks also result from *alterations* in gene flow such as *decreased* migration, *expansion* into new habitats, or population s*ubdivision*.

Natural Selection

Natural selection is the process by which genetic mutations that enhance reproduction become, and remain, more common in successive generations of a population. It has been called a "self-evident" mechanism because it necessarily follows from these simple facts:

* Heritable variation exists within populations of organisms.

* Organisms produce more offspring than can survive.

* Their offspring vary in their ability to survive and reproduce.

These conditions produce *competition* between organisms for survival and reproduction.

Consequently, organisms with traits that give them an advantage over their competitors pass these advantageous traits on, while traits that do not confer an advantage are not p*assed* on to the next generation.

The control concept of natural selection is the evolutionary *fitness* of an organism.

This measures the organism's genetic contribution to the next generation. This, however, is *not* the same as the total number of offspring: instead fitness measures the proportion of subsequent generations that carry an organism's genes. Consequently, if an allele increases fitness more than the other alleles of that gene, then with each generation this allele will become more common within the population. These traits are said to be "selected **for**." Examples of traits that can increase fitness are enhanced survival, and increased fecundity. Conversely, the lower fitness caused by having a less beneficial or deleterious allele results in this allele becoming rarer – they are "selected **against**."

Importantly, the fitness of an allele is not a *fixed* characteristic, if the environment changes, previously neutral or harmful traits may become beneficial and previously beneficial traits become harmful.

Natural selection within a population for a trait that can vary across a range of values, such as height, can be categorized into three different types: *First* is *directional selection*, which is a shift in the average value of a trait over time – Organisms, for example, slowly getting taller. *Secondly*, disruptive selection is selection for extreme trait values and often results in two different values becoming most common, with selection *against* average value. This would be when either *short* or *tall* organisms had an advantage, but not those of *medium* height. And *thirdly*, in stabilizing selection there is selection *against* extreme trait values on both ends, which causes a *decrease* in variance around the *average* value. This would, for example, cause organisms to slowly become all the same height.

A special case of natural selection is *sexual selection*, which is selection for any trait that increases mating success by increasing the attractiveness of an organism to potential mates. Traits that evolved through sexual selection are particularly prominent in *males* of some animal species, despite cumbersome antlers, mating calls, or bright colors, that a*ttract* predators, decreasing the survival of individual males. This survival disadvantage is balanced off by higher reproductive success in males that show these hard to *fake*, sexually selected traits.

An active area of research is the unit of selection, with natural selection being proposed to work at the level of *genes, cells*, *individual* organisms, *groups* of organisms and even s*pecies*. None of these models are mutually exclusive, and selection may act on multiple levels simultaneously. *Below* the level of the *individual*, genes called t*ransposons* try to copy themselves throughout the genome. Selection at a level *above* the i*ndividual*, such as group selection, may allow the evolution of cooperation, as discussed below.

Genetic drift

Genetic drift is the *change* in allele frequency from one generation to the next that occurs because alleles in offspring are a *random sample* of those in the parents, as well as from the role that *chance* plays in determining whether a given individual will survive and reproduce. In mathematical terms, alleles are subject to *sampling error*. As a result, when selective forces are absent or relatively weak, allele frequency tend to "drift" *upward* or d*ownward* (in a random walk) randomly. This population halts when an allele eventually becomes *fixed*, either by disappearing from the population, or replacing the other alleles entirely. Genetic drift may therefore *eliminate* some alleles from the population due to c*hance* alone. Even in the absence of selective forces, genetic drift can cause two separate populations which began with the same genetic structure to drift into *two* divergent populations with different sets of alleles.

The time for an allele to become fixed by genetic drift depends on population *size*, with fixation occurring more *rapidly* in *smaller* populations. The precise measure of population that is important here is called the *effective* population size, which was defined by *Sewall Wright* as a theoretical number representing the number of breeding individuals that would exhibit the same degree of breeding.

Although natural selection is responsible for *adaptation*, the relative importance of the two forces of *natural selection* and *genetic drift* in driving evolutionary change in general, is an area of current research in evolutionary biology. These investigations were prompted by the neutral theory of *molecular* evolution, which proposed that most evolutionary changes are the result of the *fixation of neutral mutations* that do not have any immediate effects on the fitness of the organism. Hence, in this model, most genetic changes in a population are the result of constant mutation pressure and genetic drift.

Gene flow

Gene flow is the *exchange* of genes between populations, which are usually of the same species. Examples of gene flow within a species include the migration and then breeding of organisms, or the exchange of *pollen*. Gene transfer between species includes the formation of *hybrid* organisms and *horizontal* gene transfer.

Migration into or out of a population can change allele frequencies, as well as introducing genetic variation into a population. Conversely, *emigration* may remove genetic material. As barriers to reproduction between two diverging populations are required for populations to become new species, gene flow may slow this process by spreading genetic differences between the populations. Gene flow is hindered by m*ountain ranges, oceans* and *deserts* or even man-made structures such as the *Great Wall* of China, which has hindered the flow of plant genes.

Depending on how far two species have diverged since their most recent common ancestor, it may still be possible for them to produce offspring, as with horses and donkeys mating to produce mules. Such *hybrids* are generally infertile, due to the two different sets of chromosomes being unable to pair up during **meiosis**. In this case, closely related species may regularly interbreed, but hybrids will be selected **against**, and the species will remain distinct. Viable hybrids, however, are occasionally formed and these *new* species can either have properties intermediate between their parent species, or possess a totally *new* phenotype. The importance of hybridization in creating *new* species of animals is *unclear*, although cases have been seen in many types of animals, with the gray tree frog being a particularly well-studied example.

Hybridization is, however, an important means of speciation in plants, since p*olyploidy* (having more than *two* copies of each chromosome) is tolerated in *plants* more than in *animals*. Polyploidy is important in hybrids as it allows reproduction, with the *two* different sets of chromosomes, each being able to *pair* with an identical partner during meiosis. *Polyploids* also have *more* genetic diversity, which allows them to avoid interbreeding depression in small populations.

Horizontal gene transfer is the transfer of genetic material from one organism to another organism that is *not* its offspring – this is most common among *bacteria*. In medicine, this contributes of antibiotic resistance, as when one bacteria acquires r*esistance* genes it can rapidly transfer them to *other* species. *Horizontal* transfer of genes from bacteria *eukaryotes* such as yeast *Saccharomyces cerevisiac* and the *adzuki* bean beetle. *Callosobruchus chinensis* may also have occurred. An example of larger-scale transfers are the eukaryotic *belloid rotifers*, which appear to have received a range of genes from bacteria, fungi, and plants. *Viruses* can also carry DNA between organisms, allowing transfer of genes across biological domains. Gene transfer has also occurred between the ancestors of *eukaryotic* cells and *prokaryotes*, during the acquisition of c*hloraplants* and *mitochonrial*.

Outcomes

Evolution influences every aspect of the form and behavior of organisms. Most prominent are the specific behavioral and physical adaptations that are the outcome of natural selection. These adaptations *increase* fitness by aiding activities such as finding food, avoiding predators or attracting mates. Organisms can also respond to selection by c*ooperating* with one another, usually by *aiding* their relatives or engaging in mutually-beneficial *symbiosis*. In the larger term, evolution produces *new* species through *splitting* ancestral populations of organisms into *new* groups that *cannot* or *will not* interbreed.

These outcomes of evolution are sometimes divided into *macroevolution*, which is evolution that occurs *at* or *above* the level of species, such as *speciation,* and m*icroevolution*, which is smaller evolutionary changes, such as *adaptations*, within a species or population. In general, macroevolution is the outcome of long periods of microevolution. Thus, the distinction between micro and macroevolution is *not* a fundamental one – the difference is simply the *time* involved. In macroevolution, however, the traits of the entire species are important. A large amount of variation among individuals, for instance, allows a species to adapt *rapidly* to *new* habitats, l*essoning* the chance of it going extinct, while a wide geographic range *increases* the chance of *speciation*, by making it more likely that part of the population will become i*solated*. In this sense, microevolution and macroevolution can sometimes be separate.

A common *misconception* is that evolution is that it is "progressive," but natural selection has no *long-term* goal and does not necessarily produce greater complexity.

Although complex species have evolved, they occurs as a *side effect* of the overall number of organisms increasing, and *simple* forms of life remain more *common*. The overwhelming *regularity* of species, for example, are microscopic *prokaryotes*, which form about *half* of the world's *biomass* despite their *small* size, and constitute the *vast* majority of Earth's biodiversity. Simple organisms have therefore been the dominant forms of life on Earth throughout its history and continue to be the main form of life to the present day, with complex life only appearing more diverse because it is more noticeable.

Adaptation

Adaptations are structures or behaviors that enhance a specific function, causing organizations to become *better* at surviving and reproducing. They are produced by a combination of the continuous production of *small, random* changes in traits, followed by natural selection of the variants best-suited for their environment. This process can cause either the *gain* of a new feature, or the *loss* of an ancestral feature. An example that shows both types of change is bacterial adaptation to antibiotic selection, with mutations causing antibiotic resistance by either *modifying* the target of a drug, or the *removing* of transporters that allow the drug into the cell. Other striking examples are the bacteria *Escherichia coli* evolving the ability to use citric acid as a nutrient in a long-term laboratory experiment, or *Flavobacterium* evolving a novel enzyme that allows these bacteria to grow on the byproducts of nylon manufacturing.

Many traits, however, that appear to be simple adaptations are in fact *exaptations*, structures originally adapted for *one* function, but which coincidentally became somewhat useful for some *other* functioning the process. One example is the African lizard *Holapsis guenthert*, which developed an extremely flat head for hiding in crevices, as can be seen by looking at its near relatives. In this species, however, the head has become so *flattened* that it assists in gliding from tree to tree – an *exaptation*. Another is the recruitment of enzymes from *glycolysis* and *xenobiotic* metabolism to serve as structural proteins called *crystallins* within the lenses of organisms' eyes.

An adaptation occurs through the gradual modification of existing structures, structures with similar internal organization, may have very different functions in related organisms. This is the result of a single ancestral structure being adapted to function in different ways. The bones within bat wings, for example, are structurally similar to both h*uman hands* and *seal flippers*, due to the common descent of these structures from an a*ncestor* that had *five* digits at the end of each forelimb. Other idiosyncratic anatomical features, such as bones in the wrist of the panda being formed into a *false* "thumb," indicate that an organism's evolutionary lineage can limit what adaptations are possible.

During adaptation, some structures may *lose* their original function and become v*estigial* structures. Such structures may have *little or no* function in *current* species, yet have had a clear function in *ancestral* species, or other closely-related species. Examples include *pseudo-genes*, the non-functional remains of eyes in blind *cave-dwelling* fish, wings in *flightless* birds, and the presence of *hip bones* in whales and snakes. Examples of *vestigial* structures in humans include *wisdom teeth,* the *coccyx*, and the *vermiform appendix*.

An area of current investigation in evolutionary developmental biology is the development basis of adaptations and *exaptations*. This research addresses the *origin* and *evolution* of embryonic development and how modifications of development and developmental processes produce novel features. These studies have shown that evolution can *alter* development to *create* new structures, such as embryonic bone structures that have been lost in other animals instead forming part of the middle ear in mammals. It is also possible for structures that have been *lost* in evolution to *reappear* due to changes in developmental genes, such as a mutation in chickens causing embryos to grow teeth similar to those of crocodiles.

Interaction between organisms can produce both *conflict* and *co-operation*. What the interaction is between pairs of species such as a *pathogen* and a *host*, or a *predator* and its *prey*, those species can develop matched sets of adaptations. Here, the evolution of one species causes adaptation in a second species. These changes in the *second* species then, in turn, cause *new* adaptations in the first species. This cycle of selection is called *co-evolution*. An example is the production *tetrodtoxin* in the rough-skinned *newt* and the evolution of *tetrodotoxin* resistance in its predator, the *garter snake*. In this predator-prey pair, an evolutionary arms race has produced *high* levels of *toxin* in the newt and correspondingly *high* levels of *resistance* in the snake.

Co-evolution

Interaction between organisms can produce both *conflict* and *co-operation*. When the interaction is between pairs of species, such as pathogen and a host, or a predator and its prey, these species can develop matched sets of adaptations. Here, the evolution of one species causes adaptation in a second specie. These changes in the *second* species then, in turn, cause *new* adaptations in the first species. This cycle of selection and response is called *co-evolution*. An example is the production of *tetrodotoxin* in the rough-skinned newt and the evolution of tetrodotoxin in its predator, the common garter snake. In this predator-prey pair, an evolutionary arms race has produced high levels of *toxin* in the newt and correspondingly high levels of *resistance* in the snake.

Co-operation

Not *all* interactions between species, however, involve *conflict*. Many cases of mutually beneficial interactions have evolved. An extreme cooperation, for instance, exists between plants and the *mycorrhizal* fungi that grow on their roots and aid the plant in absorbing nutrients from the soil. This is a *reciprocal* relationship as the plants provide the fungi with sugars from *photosynthesis*. Here, the fungi actually grow *inside* the plant cells, allowing them to *exchange* nutrients with their *hosts*, while sending signals that *suppress* the plant's immune system.

Coalitions between organisms of the same species have also evolved. An extreme case is the *eussociality* found in social insects, such as *bees, termites* and *ants*, where sterile insects feed and guard the small number of organisms in a colony that are able to reproduce. On an even smaller scale, the somatic cells that make up the body of an animal *limit* their reproduction so they can maintain a *stable* organism, which then supports a small number of the animal's germ cells to produce offspring. Here, somatic cells respond to *specific signals* that instruct them to either *grow* or *kill* themselves. If cells *ignore* these signals and attempt to multiply *inappropriately*, their *uncontrolled* growth causes *cancer*.

These examples of co-operation within species are thought to have evolved through the process of *kin selection*, which is where one organism acts to help raise a relative's offspring. This activity is selected *for* because if the *helping* individual contains alleles which promote the *helping* activity, it is likely that its kin will *also* contain these alleles and thus theses alleles will be *passed* on. Other processes that may promote cooperation include *group* selection, where co-operation provides benefits to a group of organisms.

Speciation

Speciation is the process where a species diverges into *two or more* descendent species, where it has been observed multiple times under both controlled laboratory conditions and in nature. In sexually reproducing organisms, speciation results from *reproductive isolation* followed by *genealogical divergence*. There are four mechanisms for speciation: (1) *allopatric*, (2) *periparric*, (3) *parapatric,* (4) *sympatric*. The most common in animals is *allopatric* speciation, speciation which occurs in populations initially isolated geographically, such as by habitat fragmentation or migration. Selection under these conditions can produce *very rapid* changes in the appearance and behavior of organisms. As *selection* and *drift* act independently on populations *isolated* from the rest of their species, speciation may eventually produce organisms that *cannot* interbreed. The second mechanism of speciation is *periparric* speciation, which occurs when small populations of organisms become isolated in a new environment. This differs from a*llopatric* speciation in that the isolated populations are numerically much *smaller* than the *parental* population. Here, the *founder effect* causes rapid speciation through both rapid genetic drift and selection on a small scale.

The third mechanism of speciation is *parapatric* speciation. This is similar to p*eripatric* speciation in that a small population enters a new habitat, but differs in that there is no physical separation between these two populations. Instead, speciation results from the evolution of mechanisms that *reduce* gene flow between the two populations. This generally occurs when there has been a *drastic change* in the environment within the parental species habitat. One example is the grass *Anthoxanthum odurohum,* which can undergo *parapatric* speciation in response to localized metal pollution from mines. Here, plants evolve that have resistance to high levels of metals in the soil. Selection *against* interbreeding with the metal-sensitive parental population produces a change in flowering time of the metal-resistant plants, causing reproductive isolation. Selection *against* hybrids between the two population may cause *reinforcement*, which is the evolution of traits that promote mating within a species, as well as character displacement, which is when two species become more distinct in appearance.

Finally, in *sympatric* speciation *sepias* diverge without geographic isolation or change in habitat. This form is rare since even a *small* amount of gene flow may remove genetic d*ifferences* between parts of a population. Generally, a *sympatric speciation* in animals requires the evolution of *both* genetic differences and *non-random* mating, to allow reproduction isolation to evolve.

One type of sympatric speciation involves *cross-breeding* of two related species to produce a *new* hybrid species. This is *not* common in animals as *animal* hybrids are usually sterile, because during *meiosis* the *homologous chromosomes* from each parent, being from *different species cannot* successfully *pair*. It is more common in plants, however, because plants often *double* their number of chromosomes, to form *polyploids*.

This allows the chromosomes from each parental species to form a *matching pair* during m*eiosis*, as each parent's chromosomes is represented by a pair already. An example of such a speciation is when the plant species *Arabidpsis thaliana* and *Arabidopsis arenosa* cross-bred to get the new species *Arabidopsis suecica*. This happened about 20,000 years ago, and the speciation process has been repeated in the laboratory which allows the study of the genetic mechanisms involved in the process. Indeed, chromosomes doubling within a species may be a common cause of reproductive isolation, so half the doubled chromosomes will be unmatched when breeding with un-doubled organisms.

Speciation events are important in the theory of punctual *equilibrium,* which accounts for the pattern in the fossil record of short "bursts" of evolution interspersed with relatively long periods of *stasis*, where species remain relatively unchanged. In this theory, *speciation* and *rapid* evolution are linked, with natural selection and genetic drift acting most strongly on organisms undergoing speciation in novel habitats or small populations. As a result, the periods of stasis in the fossil record correspond to the p*arental* population, and the organisms undergoing speciation and rapid evolution are found in small populations or geographically-restricted habitats, and therefore rarely being preserved as fossils.

Extinction

Extinction is the disappearance of an entire species. Extinction is *not* an unusual event, as species regularly *appear* through speciation, and *disappear* through extinction. Indeed, m*ost* of animal and plant species that have lived on earth are now *extinct*. These extinctions have happened continuously throughout the history of life, although the rate of extinction spikes in occasional *mass extinction* events. The *Cretaceous-Tertiary* extinction event, during which the dinosaurs went extinct, is the most well-known, but the earlier *Permian-Triassic* extinction event was even more *severe*, with approximately **96 percent** of species driven to extinction. The **Holocene extinction** event is an ongoing mass extinction associated with *humanity's* expansion across the globe over the past few thousand years. Present-day extinction rates are **100** *to***1000** times greater than the background rate, and up to **30 percent** of species may be extinct by the mid-21st century. *Human activities* are now the primary cause of the ongoing extinction event, *global warming* may further accelerate it in the future.

The *role* of extinction in evolution depends on which type is considered. The causes of the continuous "low-level" extinction events, which form the *majority* of extinctions, are not well understood and may be the result of *competition* between species for shared resources. If competition from other species does alter the probability that a species will become extinct, this could produce species selection as a level of natural selection. The i*ntermittent* mass extinctions are also important as a *selective* force, they drastically r*educe* diversity in a nonspecific manner and promote bursts of rapid evolution and speciation to survivors.

INTRODUCING EVOLUTIONARY HISTORY OF LIFE

Origin of life

The origin of life is a *necessary* precursor for biological evolution, but understanding that evolution occurred once organisms *appeared*, and investigating how this happens, does *not* depend on understanding exactly how life began. The current scientific consensus is that the complex biochemistry that makes up life came from *simpler* chemical reactions, but it is *unclear* how this occurred. *Not* much is certain about the *earliest* developments in life, the structure of the *first* living things, or the identity and nature of any *last* universal common ancestor or ancestral gene pool. Consequently, there is *no* scientific consensus on how life *began*, but proposals include *self-replicating* molecules such as *RNA*, and the assembly of simple cells.

Common descent

All organisms on Earth are descended from a *common ancestor* or *ancestral gene pool*.

Current species are a stage in the process of a long series of *speciation* and *extinction* events. The common descent of organisms was *first* deduced from *four* simple facts about organisms:

First: they have geographic distributions that cannot be explained by local adaptations.

* *Second:* the diversity of life is not a set of completely unique organisms, but organisms that share morphological similarities.
* *Third:* vestigial traits with no purpose resemble functional ancestral traits, and
* *Finally:* that organisms can be classified using similarities into a hierarchy of nested groups.

Past species have also left records of their evolutionary history. Fossils, along with competitive anatomy of present-day organisms, constitute the morphological, or anatomical, record. By comparing the anatomies of both modern and extinct species, paleontologists can *infer* the lineages of these species. This approach, however, is the most successful for *only* organisms that have had *body parts*, such as *shells, bones* or *teeth*. Further, as *prokaryotes* such as bacteria and *archaea* share a limited set of common morphologies, their fossils do not provide information on their ancestry.

More recently, evidence for common descent has come from the study of biochemical *similarities* between organisms. All *living* cells, for example, use the same *nucleic* acids and *amino* acids. The *development* of molecular genetics has revealed the record of evolution left in organisms' genomes: dating when species *diverged* through the molecular clock produced by *mutations*. These DNA sequence comparisons, for example, have revealed the *close* genetic similarity between *humans* and *chimpanzees* and shed *new* light on when the common ancestor of these species existed.

Evolution of life

Despite the *uncertainty* on how life began, it is clear that *prokaryotes* were the first organisms to inhabit the Earth, approximately *3-4 billion years ago*. No obvious changes in morphology or cellular organization occurred in these organisms over the next few billion years.

The *eukaryotes* were the *next* major innovation in evolution. These came from ancient bacteria being *engulfed* by the ancestors of eukaryotic *celles*, in a cooperative association called *endosymbiosis*. The engulfed bacteria and the hoist cell then underwent co- evolution into either *mitochondria* or *hydrogenosomes*. An independent second engulfment of *evanobacterial-like* organisms led to the formation of *chloroplasts* in algae and plants.

The history of life was that of the unicellular *eukaryotes*, *prokaryotes*, and *archaea* until about a *billion* years ago when multi-cellular organisms *began* to appear in the oceans in the *Ediacaron* period. The evolution of *multi-cellularity* occurred in *multiple independent events*, in organisms as diverse as *sponges, brown algae, cyanobacteria, slime moulds* and *myxobacteria*.

Soon after the emergence of these first multi-cellular organisms, a remarkable amount of biological diversity appeared over an approximate **10 million year** span, in an event called the **Cambrian explosion**. Here, the majority of types of modern animals appeared in the fossil record, as well as unique lineages that subsequently became extinct. Various triggers for the Cambrian explosion have been proposed, including the *accumulation* of *oxygen* in the atmosphere from *photosynthesis*. About *500* million years ago, *plants* and *fungi* colonized the land, and were soon followed by *arthropods* and other animals.

Amphibians first appeared around **300 million years ago**, followed by early *amniotes*, That include the *Synapsida* (mammals and mammal-like reptiles) and *Sauropsida* (reptiles and dinosaurs) then mammals around *200* million years ago, and birds around **100 million years ago** (both from "reptile-like" lineages) Despite the evolution of these large animals, however, smaller organisms similar to the types that evolved early in the process continue to be highly successful and *dominate* the Earth, with the majority of both *biomass* and *species* being *prokaryotes*.

History of evolutionary thought

Evolutionary ideas such as common descent and the transmutation of species have existed since at least the **6th century BC**, when they were expounded by Greek philosopher *Anaximander*. Others who considered such ideas included the Greek philosopher *Empedocles*, the Roman philosopher-poet *Lucretius*, the Arab biologist *Al-Jahiz*, the Persian philosopher *Miskawayh*, the *Brethren of Purity*, and the Eastern philosopher *Zhuangzi*. As biological knowledge grew in the **18th century**, evolutionary ideas were set out by a few natural philosophers including *Pierre Maupertuis* in **1745** and *Erasmus Darwin* in **1796**. The ideas of biologist *Jean-Baptiste Lamarck* about *transmutation of species* had wide influence. **Charles Darwin** formalized his idea of **natural selection** in **1838** and was still developing his theory in **1858** when *Alfred Russell Wallace* sent him a similar theory, and both were presented to the *Linnean Society of London* in separate papers. At the end of 1859 Darwin's publication of **On the Origin of the Species** explained natural selection in detail and presented evidence leading to increasingly *wide acceptance* on the *occurrence* of evolution.

Debate about the *mechanisms* of evolution continued, and still continues, but Darwin, however, could *not* explain the source of *heritable variations* which could be *acted* on by *natural selection*. Like *Lamarck*, he thought that parents passed on adaptations acquired during their lifetimes, a theory which was subsequently dubbed *Lamarckism*. In the 1880s, *August Weismann's* experiments indicated that changes from use and disuse were not heritable, and Lamarckism gradually fell from favor. More significantly, Darwin could *not* account for how traits were passed down from *generation* to *generation*. In **1865**, **Gregor Mendel**, a Gregorian monk, found that *traits* were *inherited* in a *predictable* manner. When Mendel's work was *rediscovered* in **1900**, disagreements over the *rate* of evolution predicted by *early* geneticists and biometricians, led to a rift between the *Mendelian* and *Darwinian* models of evolution.

This *contradiction* was *reconciled* in the **1930s** by biologists such as *Ronald Fisher*.

The end result was a *combination* of evolution by *natural selection* and *Mendelian inheritance*, the *modern* evolutionary synthesis. In the **1940s**, the identification of *DNA* as the genetic material by *Oswald Avery* and colleagues and the subsequent publication of the S*tructure of DNA* by **James Watson** and **Francis Crick** in **1953**, demonstrated the *physical* basis for inheritance. Since then *genetics* and *molecular* biology have become *core* parts of *evolutionary biology* and have *revolutionized* the field of p*hylogenetics.* In its early history, evolutionary biology *drew* primarily on scientists from traditional t*axonomically-oriented* disciplines, whose specialist training in particular organisms addressed general questions in evolution. As evolutionary biology *expanded* as an a*cademic discipline*, particularly *after* the development of modern evolutionary synthesis, it began to *draw* more widely from the *biological* sciences. Currently the study of evolutionary biology involves scientists from fields as diverse as *biochemistry, ecology, genetics* and *physiology*, and evolutionary concepts are used in even more distant disciplines such as *psychology, medicine, philosophy* and *computer science.*

Social and cultural responses

Even before the publication of *On the Origins of Species*, the idea that *life* had *evolved* was an *active* source of debate. Evolution is still a *contentious* concept in some quarters o*utside* the *scientific* community. Debate had centered on the *philosophical, social* and r*eligious* implications of evolution, *not* on the science itself – the proposition that biological evolution *occurs* through the *mechanism* of natural selection is standard literature.

Although many *religious denominations* have *reconciled* their beliefs with evolution through various concepts of *theistic evolution*, there are many *creationists* (*intelligent design),* who believe that evolution is *contradicted* by the creation myths found in their respective religions. As Darwin recognized early on, the *most* controversial aspect of evolutionary thought is its implications for *human origins*. In some countries – notably the United States – these *tensions* between *scientific* and *religious* teachings have fueled the ongoing *creation-evolution* controversy, a religious conflict focusing on politics and public education. While other scientific fields such as *cosmology* and *earth science* also conflict with interpretations of many religious texts, *evolutionary biology* experiences significantly more *opposition* from many religious believers.

Evolution has been used to *support* philosophical positions that promote d*iscrimination* and *racism.* For example, the *eugenic ideas* of *Francis Galton* were developed to argue that the human *gene pool* should be *improved* by *selective breeding* policies, including *incentives* for those considered "good stock" to reproduce, and the c*ompulsory* sterilization, *prenatal* testing, *birth* control, and even *killing*, of those considered "bad stock." Another example of an *extension* of evolutionary theory that is now widely regarded as *unwarranted* is "Social Darwinism," a term given to the 19th century *Whig Malthusian theory* developed by *Herbert Spencer* into ideas about "survival of the fittest" in commerce and human societies as a whole, and by others into claims that s*ocial inequity, racism*, and *imperialism* were *justified.* Contemporary scientists and philosophers, however, consider these ideas to have been *neither* mandated by evolutionary theory *nor* supported by *data.*

Applications

A *major* technological application of evolution is *artificial selection,* which is the i*ntentional* selection of certain traits in a population of organisms. Humans have *used* a*rtificial* selection for thousands of years in the *domestication* of *plants* and *animals.*

More recently, such selection has become a vital part of *genetic engineering*, with selectable markers such as antibiotic resistance genes being used to *manipulate* DNA in molecular biology.

As evolution can produce *highly optimized* processes and networks, it has many a*pplications* in computer science. Here, *simulation* of evolution using evolutionary a*lgorithms* and *artificial life* started with the work of *Nils Aall Barriclli* in the **1960s**, and was extended by *Alex Fraser*, who published a series of papers on simulation of artificial selection. *Artificial evolution* became a widely recognized optimization method as a result of the work of *Ingo Rechenberg* in the **1966s** and early **1970s**, who used evolution strategies to solve complex engineering problems. *Genetic algorithms* in particular became *popular* through the writing of *John Holland*. As *academic* interest grew, dramatic increases in the power of computers allowed practical applications, including the *automatic* evolution of computer programs. *Evolutionary algorithms* are now used to solve *multi-dimensional* problems *more efficiently* than *software* produced by *human* designers, and also *optimize* the design of systems.

Understanding *macroevolution* can have *practical* applications too. A certain species of coral might be *discovered* that produces an *antibiotic* with medical potential. Knowing its *closest* relatives would inform researchers of *other* species that might produce *similar* compounds, which could then be investigated.

Biologists studying evolution do a variety of things: *population geneticists* study the p*rocess* as it is occurring – *systematists* seek to determine *relationships* between species and *paleontologists* seek to uncover details of the *unfolding* of life in the past. Discerning these details is often difficult, but *hypotheses scientists* have, concerning the history of the planet.

The material here ranges from some issues that are *fairly* certain to some topics that are nothing more than *informed* speculation. For some points, there are *opposing* hypotheses.

A Brief History of Life

Life evolved in the sea. It stayed there for a majority of the history of the earth.
The first *self-replicating* molecules were most likely *RNA*. *RNA* is a *nucleic acid* similar to *DNA*. In laboratory studies it has been shown that RNA sequences have *catalytic* capabilities. Most importantly, certain RNA sequences act as *polymerases* – enzymes that form *strands* of RNA from its *monomers*. This process of *self-replication* is the *crucial step* in the *formation* of *life*, and is called the *RNA world hypothesis*.

The *common ancestor* of all life doubtlessly used RNA as its genetic material. This ancestor gave rise to *three major lineages* of life. These are: (1) the *prokaryotes* ("ordinary" bacteria), (2) *archaebacteria* (*thermophilic, methanogenic* and *halophilic* bacteria) and (3) *eukaryoyes*. Eukaryote include *protests* (single-celled organisms like a*moebas* and *diatoms* and a few multi-cellular forms such as *kelp*), *fungi* (including y*east*), plants and animals. *Eukaryotes* and *archaebacteria* are the two most *closely related* of the three. The process of *translation* (making *proteins* from the instructions on a messenger RNA template) is *similar* in these lineages, but the *organization* of the genome and *transcription* (making messenger RNA from a DNA template) is very d*ifferent* in prokaryotes than in eukaryotes and archbacteria. Scientists interpret this to mean that the *common ancestor* was RNA based; it gave rise to *two lineages* that independently formed a DNA genome and hence *independently* evolved mechanisms to t*ranscribe* RNA into DNA.

The *first* cells must have been *anacerobic* because there was *no* oxygen in the a*tmosphere*. In addition, they probably were **thermophlic ("heat-loving")** and f*ermentive*. Rocks as old as **3.5 billion years old** have yielded *prokaryotic* fossils.

Specifically, some rocks from Australia called the *Warrawoona series* give evidence of bacterial communities organized into structures called **stromatolites**. Fossils like these have subsequently been *found* all over the world. These *mats* of *bacteria* still *form* today in a few locales (Spark Bay Australia, for example). Bacteria are the only life forms found in very *old* rocks ("rock of ages"), *eukaryotes* (protests) appear about **1.5 billion years ago**, and *fungi-like things* appear about **900 million years ago**, or. **09 billion**.

Photosynthesis evolved around **3.4 billion years ago** a process that allows organisms to *harness* sunlight to *manufacture* sugar from simpler precursors. The first photo- synthesis to evolve, PSI, uses light to convert *carbon dioxide* (CO_2) and hydrogen sulfide (H_2S) to *glucose*. This process releases *sulfur* as a *waste* product.

About a **billion years** later, a *second* photosynthesis (PS) evolved, probably from a *duplication* of the first photosynthesis. Organisms with PSH use both *photosynthesis* in conjunction to convert *carbon dioxide* (CO) and *water* (H2O) into *glucose*. This process *releases* oxygen as a *waste* product. *Anoxygenic* (or H25) photosynthesis, using PSI, is seen in *living* purple and *green* bacteria. Oxygenic (or H20) photosynthesis, using PSI, takes place in *cyanobacteria*. Cyanobacteria are *closely* related to and hence probably *evolved* from *purple* bacterial ancestors. Green bacteria are an *out-group*. Since *oxygenic* bacteria are a lineage within a cluster of *anoxygenic* lineages, scientists infer that PSI evolved *first*.

This also corroborates with geological evidence.

Both *Green plants* and *algae* also use *photosynthesis*. In these organisms, photosynthesis occurs in *organelles* (membrane-bound structures within the cell) called *chloroplasts*. These organelles originated as *free living* bacteria related to the *cyanobacteria* that were engulfed by *ur-eukaryotes* and eventually entered into an *endosymbiotic* relationship. This *endosymbiotic theory* of eukaryotic organelles was championed by *Lynn Margulis*. Originally *controversial*, this theory is now *accepted*.

One key line of evidence in support of this idea came when the DNA inside chloroplasts Was *sequenced* – the gene sequences were more similar to free-living cyanobacteria sequences than to sequences from the plants that chloroplasts *resided* in. After the advent of *photosynthesis II*, oxygen *increased*. Dissolved oxygen in the oceans *increased* as well as atmosphere oxygen. This is sometimes called the **oxygen holocaust**. Oxygen is a very good *electron acceptor* and can be very *damaging* to living organisms. Many bacteria are *anaerobic* and *die* almost immediately in the presence of oxygen. Other organisms, like animals, have special ways to *avoid* cellular damage due to this element (and in fact *require* it to live.) Initially, when oxygen began building up in the environment, it was *neutralized* by materials already present. *Iron,* which existed in high concentrations in the sea was *oxidized* and *precipitated*. Evidence of this can be seen in *banded* iron formation from this, layers of iron *deposited* on the *sea floor*. As one geologist put it, "**the world rusted**." Eventually, it grew to *high* enough concentration to be *dangerous* to living things. In *response*, many species went *extinct*, but some continued (and still continue) to thrive in *anaerobic microenvironments* and several lineages *independently* evolved oxygen respiration.

The *purple* bacteria evolved oxygen respiration by *reversing* the flow of molecules through their *carbon fixing* pathways and *modifying* their electron transport chains.

Purple bacteria also enabled the *eukaryotic* lineage to become *aerobic*. Eukaryotic cells have *membrane-bound* organisms called *mitochrondria* that take care of *respiration* for the cell. These are *endosymbionts* as are *chloroplasts*. Mitochondria formed this *symbiotic* relationship very early in eukaryotic history, and all but a few groups of eukaryotic cells have *mitichondra*. Later, a few lineages picked up choloroplasts.

Chloroplasts have *multiple* origins. Red algae picked up ur-chloroplasts from the cyanobacteria lineage. Green algae, the group *plants* evolved from, picked up different ur-chlorophyte, a lineage closely related to cyanobacteria.

Animals start appearing prior to the *Cambrian*, about **600 million years ago**. The first animals dating just before the Cambrian were found in *rocks* near Adelaide, Australia.

They are called the *Ediacarian fauna* and have subsequently been found in other locales as well. It is unclear if these forms have any surviving descendents. Some took a bit like *Cnidarians* (jellyfish, sea anemones and the like); others resemble *annelids* (earthworms). All the *phyla* (the second highest taxonomic category) of animals appeared around the Cambrian. The Cambrian 'explosion' may have been a result of *higher* oxygen concentrations enabling *larger* organisms with *higher* metabolism to evolve. Or it might be due to the *spreading* of shallow seas at that time providing a variety of *new* niches. In any case, this *radiation* produced a wide variety of animals.

Some paleontologists think more animal *phyla* were present *then* than now. The animals of the *Burgess shale* are an example of Cambrian animal fossils, discovered in Canada, and show a *bizarre* array of creatures, some which appear to have *unique* body plans *unlike* those in any modern animals.

The extent of the Cambrian explosion is often *overstated.* Although *rapid*, the Cambrian explosion is *not* considered *instantaneous* in geological time. There is evidence also of animal life *prior* to the Cambrian. Although all the phyla of animals came into being, they were *not* the *modern forms* of today. Our own *phylum* (that we share with other mammals, *reptiles, birds, amphibians* and *fish*) was represented by a *small*, silver-like thing called **Pikaia**. Plants were not yet present. Photosynthesized *protests* and a*lgae* were at the *bottom* of the food chain. Following the Cambrian, the number of marine families leveled off at a little less than **200**.

The *Ordovician explosion*, around **500 million years ago**, followed. This 'explosion', larger than the *Camrbian*, introduced numerous families of the Paleozoic fauna (including crinoids, articulate brachiopods, cophalopods and corals). The Cambrian fauna (*tribobites, inarticulate brachiopods*, etc.) declined *slowly* during this time. By the end of the Ordovician, the Cambrian *fauna* has mostly given way to the Paleozoic *fauna* and the number of marine families was just over **400**. It stayed at this level until the end of the *Permian* period.

Plants evolved from *green* algae over **400 million years ago**. Both groups use c*hlorophyll a* and *b* as photosynthetic pigments. In addition, *plants* and green *algae* are the only groups to store *starch* in their chloroplasts. Plants and fungi (in symbiosis) invaded the land about **400 million years ago**. The first plants were *moss-like* and required *moist* environments to survive. Later, evolutionary developments such as a *waxy cuticle* allowed some plants to exploit more inland environments. Still mosses *lacked* true vascular tissue to transport fluids and nutrients. This *limited* their size since these must diffuse through the plant.

Vascular plants *evolved* from mosses. The first vascular land plant known is *Cooksonia*, a spiky branching *leafless* structure. At the same time, or shortly thereafter, a*rthropods* followed plants onto the land. The *first* land animals to come on shore were known as *myriapods – centipedes* and *millipedes.*

Vertebrates moved onto land by the *Devonian period*, about **380 million years ago**.

Ichthyostega, an *amphibian*, is among the *first* known land vertebrates. It was found in Greenland and was derived from *lobe-finned* fishes called **Rhipidistians**. *Amphibians* gave rise to *reptiles*. Reptiles had evolved *scales* to *decrease* water loss and shelled egg permitting young to be *hatched* on land. Among the earliest well preserved reptiles is *Hylonomus*, found in rocks from Nova Scotia. The **Permian extinction** was the **largest extinction** in history, which happened about **250 million years ago**. The *last* of Cambrian fauna went extinct. The *Paleozoic* fauna took a *nose dive* from about **300** families to about **50**. It is estimated that **96%** of all species (**50%** of all families) met their end. Following this event, the Modern fauna, which had been slowly *expanding* since the *Ordovician*, took over.

The modern fauna includes *fish, bivalves, gastropods* and *crabs*. These were *barely* affected by the *Permian* extinction. The Modern fauna subsequently increased to over **600** marine families at present. The *Paleozoic* fauna held steady at about **100** families. A s*econd* extinction event shortly following the Permian kept animal diversity *low* for awhile.

During the *Carboniferous* (the period just prior to the Permian) and in the Permian the landscape was dominated by *ferns* and their *relatives*. After the Permian extinction, g*ymnosperm* (ex. Pines) became more abundant. *Gymnosperms* had evolved *seeds*, from s*eedless* fern ancestor, which helped their ability to *disperse*. Gymnosperms also evolved p*ollen*, encased sperm which allowed for more out-crossing.

Dinosaurs evolved from **archosaur** reptiles – their closest *living* relatives are c*rocodiles*. One *modification* that may have been a *key* to their *success* was the evolution of an *upright stance*. *Amphibians* and *reptiles* have a *splayed* stance and walk with an undulating pattern because their *limbs* are modified from *fins*. Their gait is *modified* from the *swimming* movement of fish. Splay stance animals cannot *sustain* continued locomotion because they cannot *breath* while they *move* – their *undulating* movement *compresses* their *chest cavity*. Thus, they must *stop* every *few* steps and **breath** fresh **air** before *continuing* on their way. Dinosaurs evolved an *upright* stance similar to the *upright* mammals, which *independently* evolved. This allowed for **continued** locomotion.

Dinosaurs additionally evolved to be **warm-blooded**. Warm-bloodedness allows an *increase* in the **vigor** of movements in *erect* organisms. Splay stance organisms would probably *not* benefit from warm-bloodedness. Birds *evolved* from *sauriscian* dinosaurs. *Cladistically birds* are dinosaurs. The *transitional* fossil **Archaesopteryx** has a mixture of *reptilian* and *avian* features.

Angiosperms evolve from *gymnosperms*, whose closest relatives are *Gnetae*. *Two* key adaptations allowed them to displace gymnosperms as the dominant fauna – *fruits* and *flowers*. Fauna (modified plant ovaries) allow for animal-based seed dispersal and deposition with plenty of fertilizer. *Angiosperms* currently dominate the flora of the world – *over* **three fourths** of all living plants are **angiosperms**.

Insects evolved from primitive **segmented arthropods**. The mouth-parts of insects are *modified* legs. Insects are closely related to *annelds*. Insects *dominate* the fauna of the world. Over *half* of all named species are *insects*, **one third** of which are **beetles**.

The *end* of the *Cretaceous,* about **65 million years ago**, is marked by a *minor* mass extinction. This extinction marked the *demise* of all the lineages of *dinosaurs* save the *birds*. Up to this time, *mammals* were confined to *nocturnal, insectivorous niches*. Once the dinosaurs were out of the picture, they diversified. **Morgonucudon,** a contemporary of *dinosaurs*, is an example of one of the *first* mammals. Mammals evolved from *therapsid* reptiles. The finback reptile **Diametrodon** is an *example* of a *therapsid*. One of the most *successful* lineages of mammals is, of course, **(US) humans**. *Humans* are *neotenous* apes.

Neoteny is a process which *leads* to an organism reaching *reproductive* capacity in its juvenile form. The *primary* evidence of this are the *similarities* between young apes and *adult* humans. *Louis Bolk* compiled a list of **25 features** shared between adult humans and juvenile apes, including *facial* morphology, among which are high relative brain weight, absence of brow ridges and cranial crests.

The earth has been in a state of *flux* for **3.5 billion years**. Across this time span, the *abundance* of different lineages varies. *New* lineages *evolve* and *radiate* out across the face of the planet, pushing *older* lineages to *extinction*, or *relict* existence in *protected* refuges or *suitable* microhabitats. Organisms *modify* their environments. This can be disastrous, as in the case of the *oxygen holocaust*. Environmental *modification*, however, can be the *impetus* for further evolutionary *change*. Overall, *diversity* has *increased* since the *beginning* of life. This increase has, however, been *interrupted* numerous times by *mass* extinctions. *Diversity* appears to have *hit* an all-time high just prior to the *appearance* of humans. As human population has *increased,* biological diversity has *decreased* at an ever-increasing pace. The *correlation* is self-evidently *causal*.

INTRODUCING CELLS

Cells are the basic **building blocks** of all **living things**. The **human body** is *composed* of **trillions of cells**, which provide *structure* for the *body,* take on *nutrients* from *food*, *convert* those *nutrients* into *energy*, and carry out specialized *functions*. *Cells* also contain the *body's* **hereditary material** and can make **copies** of **themselves**.

Cells have *many* parts, *each* with a different *function*. Some of *these* parts, called **organelles**, are specialized *structures* that perform certain tasks *within* the *cell*. **Human cells** contain the following major *parts*, listed in *alphabetical order*:

Cytoplasm

Within cells, the *cytoplasm* is made up a *jelly-like* fluid (called the *cytosol*) and other structures that surround the nucleus.

Cytoskeleton

The *cytoskeleton* is a network of long *fibers* that make up the cell's *structural framework*. The *cytoskeleton* has several *critical functions*, including determining *cell shape*, participating in *cell division*, and allowing *cells* to **move**. It also *provides* a track-like *system* that *directs* the *movement* of **organelles**, along with other substances *within* cells.

Endoplasmic reticulum (ER)

This *reticulum* helps process *molecules* created by the *cell*. The *endoplasmic reticulum* also *transports* these *molecules* to their specific *destinations* either *inside* or *outside* the *cell*.

Golgi apparatus

The *Golgi apparartus* packages *molecules* processed by the *endoplasmic reticlum* to be transported *out* of the *cell*.

Lysosomes and peroxisomes

These *organelles* are the *recycling* center of the *cell*. They *digest* foreign *bacteria* that *invade* the cell, *rid* the *cell* of *toxic* substances, and recycle *worn-out* cell *components*.

Mitochondria

Mitochondria are complex *organelles* that convert *energy* from *food* into a *form* that the *cell* can *use*. *They* have their own *genetic* material, *separate* from the DNA in the *nucleus*, and can make *copies* of *themselves*.

Nucleus

The *nucleus* serves as the cell's *command center*, sending *directions* to the *cell* to *grow, mature, divide*, or *die*. It also *houses* DNA (*deoxyribonucleic acid*), the cell's *heredity* material. The *nucleus* is *surrounded* by a *membrane* called the *nuclear envelope*, which *protects* the DNA and *separates* the *nucleus* from the *rest* of the *cell*.

Plasma membrane
The *plasma membrane* is the *outer lining* of the *cell*. It *separates* the *cell* from its e*nvironment* and allows *materials* to *enter* and *leave* the *cell*.

Ribosomes
Ribosomes are *organelles* that *process* the *cell's* genetic *instructions* to create p*roteins*. These *organelles* can *float* freely in the *cytoplasm* or be *connected* to the endoplasmic *reticulum*.

The **cell** is the *structure* of all known *living* organisms. It is the *smallest* unit of an organism that is classified as **living**. *Some* organisms, such as most *bacteria*, are u*nicellular* (consisting of a *single cell*). Other organisms, such as **humans**, are m*ulticellular* – *humans* have an estimated **100 trillion cells**. A typical cell size is **10 pm;** a *typical* cell mass is **1 nanogram**. The *largest* known *cell* is in an *ostrich*. In **1837** the f*inal* cell *theory* was *developed*, by *Czech* biologist, *Jan Evangelista Purkne,* who observed small "granules" while looking at the plant *tissue* through a microscope. The c*ell theory*, first developed in **1839** by *Matthias Jakob Schleiden* and *Theodor Schwann***,** states that *all* organisms are composed of *one* or *more* cells. *All* cells come from p**re-existing** *cells*. *Vital functions* of an organism occur *within* cells, and all cells *contain* the *hereditary* information *necessary* for *regulating* cell *function* and for *transmitting* information to the *next generation* of cells.

The word **cell** comes from the Latin *cellula*, meaning, a **small room.** The descriptive name for the *smallest* living biological structure was chosen by **Robert Hooke** in a book he published in **1665** when he *compared* the *cork cells* he saw through his microscope to the **small** rooms monks **lived in**.

General principles
Each cell is at least somewhat *self-contained* and *self-maintaining*: it can *take in* n*utrients*, convert these into *energy*, carry out *specialized* functions, and *reproduce* as necessary. Each cell *stores* its own set of *instructions* for *carrying* out *each* of these a*ctivities*.

Reproduction by *cell division* is done through *binary fission/mitosis or meiosis*. The use of *enzymes* and other *proteins* for DNA *genes,* is made via *messenger* RNA i*nterdiates* and *ribosomes*. **Metabolism** includes taking in *raw materials*, building *cell* components, converting energy *molecules* and releasing *by-products*. The *functioning* of a *cell* depends upon its *ability* to *extract* and *use* chemical energy *stored* in organic m*olecules*. This energy is *released* and then used in metabolic *pathways,* that respond to *external* and *internal* stimuli, such as *changes* in temperature, *pH* or levels of *nutrients*.

Some *prokaryotic* cells contain important *internal* membrane-bound compartments.

Material is moved *between* these compartments by *regulated* traffic and the *transport* of small spheres of *membrane-bound* material called *vesicles*.

Anatomy of cells
There are two types of cells: **eukaryotic** and **prokaryotic**. *Prokaryotic* cells are usually i*ndependent*. While *eukaryotic* cells are found in *multicellular* organisms.

Prokaryotic cells
Prokaryotes cells *differ* from *eukaryotes* cells since they *lack* a *nuclear* membrane and a cell *nucleus*. Prokaryotes also lack *most* of the intracellular *organelles* and *structures* that are seen in eukaryotic cells. There are *two* kinds of prokaryotes, *bacteria* and *archaea*, but which are *similar* in the overall structures of their cells. Most functions of organelles, such as *mitochondria, chorplasts,* and the *Golgi apparatus*, are taken over by the prokaryotic cell's plasma *membrane*. Prokaryotic cells have three *architectural* regions: appendages called *flagella* and *pili* – proteins attached to the cell *surface*; a cell *envelope* consisting of a *capsule*, a cell *wall*, and a plasma

membrane, and a *cytoplasmic* region that contains the cell *genome* (DNA) and *ribosomes*, along with various sorts of *inclusions*. Other differences include:

- The plasma membrane (a phospholipid bilayer) separates the interior of the cell from its environment and serves as a *filter* and communications *beacon*.

- Most prokaryotes have a *cell wall* (some exceptions are *Mycoplasma* – bacteria) and *Thermoplasma* – archaea). This wall consists of *peptidoglycan* in bacteria, and acts as an additional *barrier* against *exterior* forces. It also *prevents* the cell from "exploding" (cytolysis) from *osmotic* pressure against a *hypotanic* environment. A cell wall is also present in some eukaryotes like *plants* (cellulose) and *fungi*, but has a *different* chemical composition.

- A prokaryotic chromosome is usually a *circular* molecule – an exception is that of the bacterium *Barrelia burgdorferi*, which causes *Lyme disease*. Even *without* a real *nucleus*, the DNA is considered as a *nucleoid*. Prokaryotes can carry *extrachromosomal* DNA elements called *plasmids*, which are usually circular. Plasmids that can carry additional functions, such as *antibiotic* resistance.

Eukaryotic cells
Eukaryotic cells are about *10 times* larger than a typical *prokaryotic* cells and can be as much as *1000 times* greater in volume. The major **difference** between *prokaryotes* and *eukaryotes* is that eukaryotic *cells* contain **membrane-bound** compart ments in which *specific* **metabolic Structure of Typical animal cell** activities take place. Most importantly among these is the presence of a **cell nucleus,** a *membrane-delineated* compartment that houses the eukaryotic cell's **DNA**.

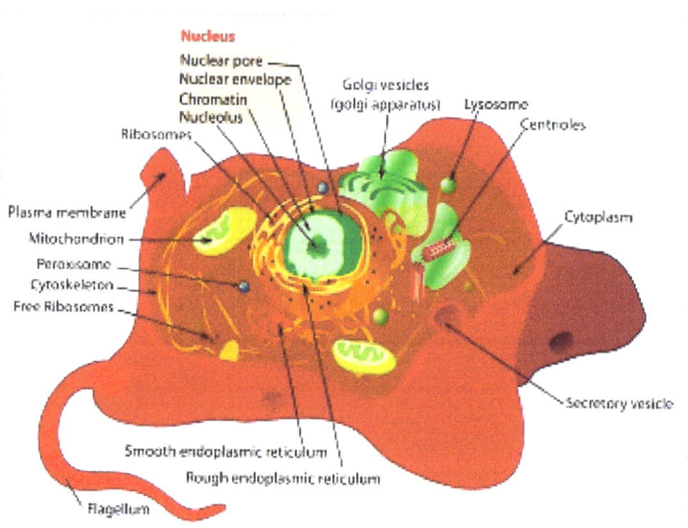

Structure of a typical animal cell.

It is *this* nucleus that gives the *eukaryotic* its name, which means "**true nucleus.**" Other differences include:

- The *plasma membrane* resembles that of *prokaryotes* in function, with minor differences in the setup. Cell walls *may* or *may not* be present.

- The eukaryotic DNA is organized in *one* or *more* linear molecules, called chromosomes, which are associated with histone proteins. All chromosomal DNA is *stored* in the cell nucleus, separated from the cytoplasm by a *membrane*. Some eukaryotic organelles such as *mitochondria* also contain some DNA.

Subcellular components
All cells, whether *prokaryotic* or *eukaryotic*, have a *membrane* that *envelopes* the cell, separates its *interior* from its *environment*, regulates what moves *in* and *out* (selectivity *permeable*), and maintain the electric potential of the cell. Inside the membrane, a salty *cytoplasma* takes up *most* of the cell volume. All cells posses **DNA**, the

hereditarymaterial of *genes*, and **RNA**, containing the *information* necessary to *build* various *proteins* such as **enzymes**, the cell's **primary** machinery. There are also other kinds of *biomolecules* in cells.

Cell membrane: A cell's defining scaffold

The *cytoplasts* of a cell is *surrounded* by a cell *membrane* or *plasma membrane*. This plasma membrane in *plants* and p*rokaryotes* is usually *covered* by a *cell* wall, which *membrane* serves to separate and protect a cell from its surrounding *environment* and is made mostly from a *double layer* of *lipids*

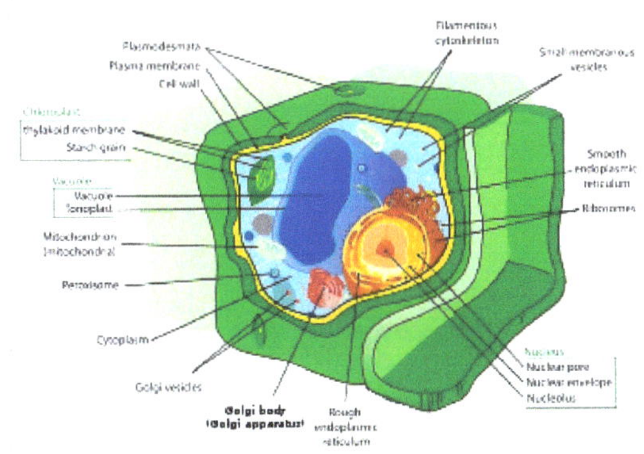

Structure of a typical plant cell.

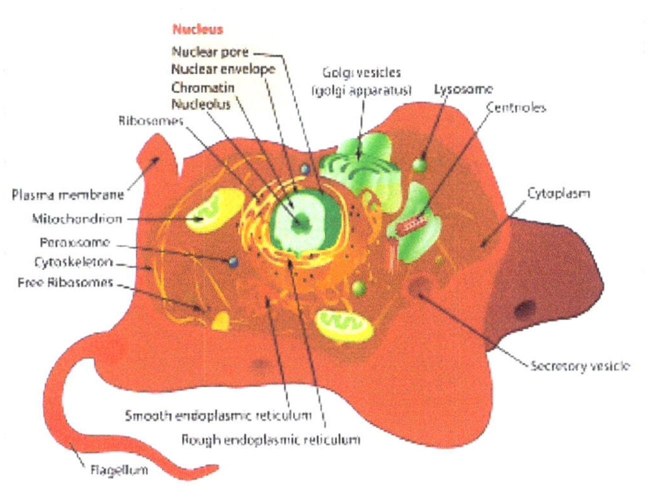

Structure of a typical animal cell.

(bydrophic fat-like molecules) and *phosphorus* molecules. Hence the layer is called a *phospholipid bilayer*. It may also be called a *fluid mosaic* membrane. Embedded within this membrane is a variety of protein molecules that act as *channels* and *pumps* that *move* different molecules *into* and *out* ofthe *cell*. The membrane is said to be'semi-permeable,' in that it can either *let* a substance (molecule orion) p*ass through* to a limited extent or *not pass* through at all. Cell surfacemembranes also contain *receptor* Phylogentic tree proteins that allow cells to direct *external*signaling molecules such as *hormones*.

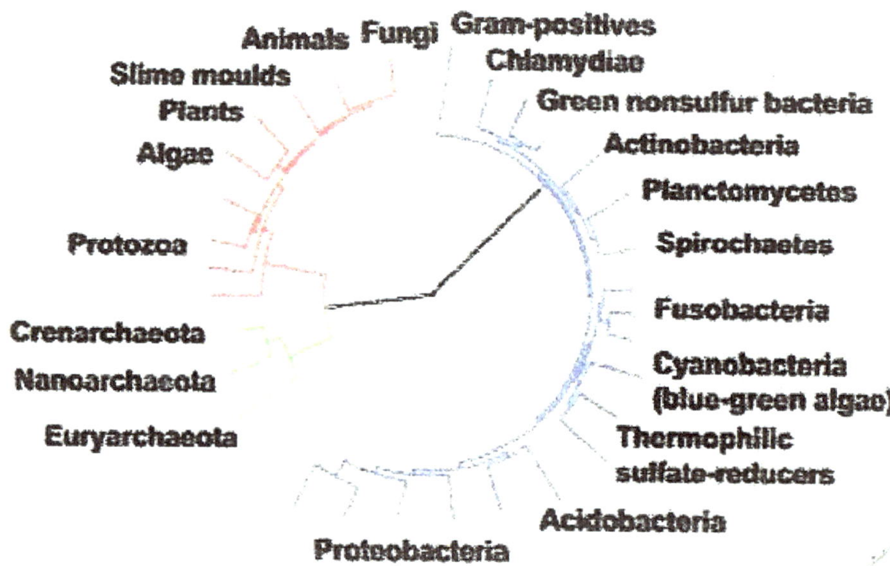

Cytoskelton: a cell's scaffold

The *cytoskelton* acts to *organize* and *maintain the* cells's *shape*; anchors *organelles* in place; helps during *endoctosis*, the uptake of external materials by a cell, and *cytokinesis*, the *separation* of *daughter* cells after *cell division*, that *moves* parts of the cell in processes of *growth* and *mobility*. The *eukaryotic cytoskelton* is composed of m*ocrofilaments*, intermediate filaments and microtubules. There is a great number of p*roteins* associated with them, each *controlling* a cell's *structure* by *directing, building, a*nd *aligning* filaments The prokaryotic *cytoskeleton* is less *well-studied* but is, nevertheless, involved in the *maintenance* of cell *shape, polarityand cytokinesis.*

Genetic material

Two different kinds of genetic material exist: *deoxyribonucleic acid* (DNA) and r*ibonucleic acid* (RNA). Most organisms use DNA for their *long-term* information storage, but some viruses (e.g., *retroviruses*) use RNA as their genetic material. The biological information contained in an organism is *encoded* in its DNA or RNA s*equence.* RNA is also used for the *genetic code* itself. *Transfer* of RNA (*tRNA*) molecules are used to *add* specific *amino acids* during the process of *protein translation.*

Prokaryotic genetic material is *organized* in a simple *circular* DNA molecule (the bacterial chromosome) in the *nucleoid* region of the *cytoplasm.* Eukaryotic genetic material is divided into *different*, linear molecules called chromosomes reside *inside* a d*iscrete* nucleus, usually with *additional* genetic material in some organelles like m*itochondria* and *chloroplasts* – see *endosymbiotic theory.*

A *human cell* has genetic material in the nucleus (the nuclear genome) that in the m*itrochondrial* genome has a circular DNA molecule *distinct* from *nuclear* DNA.

Although the Mitrochondrial DNA is *very small* compared to *nuclear chromosomes;* it e*ncodes* for **13 proteins** involved in mitrochondrial *energy* production as well as *specific* tRNAs.

Foreign genetic material (most commonly DNA) can be *artificially* introduced into the cell by a process called *transfection*. This can be *transient*, if the DNA is *not* inserted into the cell's *genome* – or if so – is *stable.* Certain viruses also *insert* their genetic material into the *genome.*

Organelles

The *human body* contains many *different* organs, such as the *heart, lung,* and *kidney*, with e*ach* organ performing *different* functions. Cells also have a *set* of "**little organs**," called o**rgamelles**, that are adapted and/or *specialized* for carrying out *one or more* **vital functions**.

There are *several* types of *organelles* within an *animal* cell. Some (such as the *nucleus* and *golgi apparatus*) are typically *solitary*, while others (such as *mitochondria, peroxisomes* and *lysosomes*) can be *numerous* (**hundreds to thousands**). The *cytosol* is the *gelatinous fluid* that *fills* the *cell* that *surrounds* the organelles.

Mitachondrial and Chloroplasts – the power generators:

* *Mitachonria* are *self-replicating* organelles that occur in various *numbers, shapes*, and s*izes* in the *cytoplasm* of all eukaryotic cells. Mitrachondria plays a *critical* role in g*enerating* energy in the eukaryotic cell. Mitachondria generates the cell's energy by the process of *oxidative phosphorylation*, utilizing *oxygen* to release energy *stored* in the cellular *nutrients* (typically pertaining to **glucose***)* to generate ATP. Mitachondria multiply by *splitting* in *two*.

* Organelles that are *modified chlorlasts* are broadly called *plastids*, that are involved in energy *storage* through the process of **photosynthesis**, which utilizes *solar* energy to generate *carbohydrates* and *oxygen* from *carbon dioxide* and *water*.

* *Mitachondria* and *chloroplasts* each contain their *own* genome, which is *separate* and d*istinct* from the *nuclear* genome of a cell. Both of these organelles contain this DNA in *circular plasmids*, much like *prokaryotic cells*, strongly supporting the evolutionary theory of **endosymbiosis** – since these organelles contain their *own* genomes and have other *similarities* to *prokaryotes*, they are thought to have been *developed* through a s*ymbiotic* relationship after being *engulfed* by a *primitive* cell.

Ribosomes

The *ribosome* in a large complex of *RNA* and *protein* molecules. This is where proteins are produced. *Ribosomes* can be found *either* floating *freely* or *bound* to a m*embrane* – the rough *endoplasmatic reticulum* in *eukaryotes*, or the cell *membrane* in prokaryotes.

Cell nucleus – a cell's information center

The *cell nucleus* is the *most conspicuous organelle* found in the *eukaryotic*, It *houses* the cell's *chromosomes*, and is the place where *almost all* DNA *replication* and RNA s*ynthesis* (transcription) occur. The nucleus is *spherical* in shape and separated from the cytoplasm by a *double membrane* called the *nuclear envelope*. The nuclear envelope *isolates* and *protects* a cell's DNA from *various* molecules that could a*ccidentally* damage its *structure* or *interfere* with its *processing*. During processing,

DNA is *transcribed*, or *copied* into a special RNA, called *mRNA*. This mRNA is then transported *out* of the nucleus, where it is *translated* into a *specific* protein molecule.

The *nucleolus* is a *specialized* region *within* the nucleus where ribosome is a*ssembled*. In prokaryotes, DNA processing takes place in the *cytoplasm*.

Endoplasmic reticulum – eukaryotes only

The *endoplasmic reticulum* (ER) is the *transport network* of molecules targeted for certain *modifications* and *specific* destinations, as compared to molecules that will f*loat* freely in the *cytoplasm*. The ER has two forms: the *rough* ER, which has r*ibosomes* on its surface that secretes *proteins* into the *cytoplasm*, and the *smooth* ER, which lacks them (proteins). Smooth ER plays a role in calcium *sequestration* and r*elease*.

Golgi apparatus – eukaryotes only

The *primary* function of the *Golgi apparatus* is to *process* and p*ackage* the *macromolecules*, such as *proteins* and *lipids* that are synthesized by the cell. It is partic- ularly *important* in the processing of proteins for *secretion*. The *Golgi* **a***pparatus* forms a part of the *endomembrane system* of eukaryotic cells. *Vesicles* that enter the *Golgi apparatus* are processed in a **cis** to *trans-direction*, meaning they *coalesce* on the cis side of the apparatus and after processing pinch off *opposite* (trans) side to force a *new* vesicle in the animal cell. Lysosomes and Peroxisomes – eukaryotes only *Lysosomes* contain *digestive* enzymes (acid hydrolases). They digest e*xcess* or *worn—out* organelles, food particles, engulfed *viruses*, and *bacteria. Peroxisomes* have *enzymes* that *rid* the cell of *toxic perodoxes*. The cell could *not* house these *destructive* enzymes if they were *not* contained in a m*embrane-bound* system. These organelles are often called a "**suicide bag**," because of their ability (if not checked), to *detonate* and *destroy* the cell.

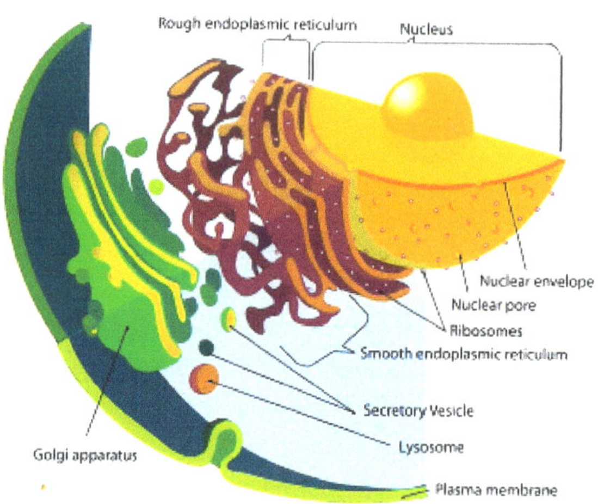

Detail of the endomembrane system and its components

Enomembrane and its components

Centrosome – the cytoskeleton organizer

The *centrosome* produces the *microtubules* of a cell – a key of the *cytoskeleton*. It d*irects* the *transport* through the ER and the Golgi apparatus. Centrososomes are c*omposed* of *two centrioles*, which *separate* during *cell division* and help in the f*ormation* of the *mitotic spindle*. A *single* centrosome is present in the *animal cell*. They are found also in some *fungi* and *algae* cells.

Vacuoles

Vacuoles store *food* and *waste*. Some vacuoles store *extra water*. They are often described as *liquid filled space* and are surrounded by a *membrane*. Some cells, most notably **Amoeba**, have *contracile* vacuoles, which are able to *pump* water *out* of the cell if there is *too much* water.

Structures outside the wall

Capsule

Capsules are present only in *some* bacteria *outside* the cell wall, and *gelatinous* in nature. The capsule may be *polysaccharide,* as in *pneumucocci, meninggococci,* or *polypeptide,* as *bacillus anthracis,* or *hyaluronic acid,* as in *streptococci*. Capsules *not* stained by o*rdinary* stain, can be detected by a *special* stain. The capsule is *antigenic*, that has a*ntiphagocytic* function to determine the *virulence* of many bacteria. It also plays a role in the *attachment* of the organism to the *mucous membrane*.

Flagella

Flagella is the *organ* of *mobility*. They *arise* from *cytoplasm* and *extrude* through the cell wall. They are *long* and *thick* thread-like *appendages,* protein in nature, and formed of *flagellin* protein (antigenic). They *cannot* be stained by *gram stain*, but have a *special* stain. According to their arrangement they may be *monotrichate, amphitrichate, lophorichate*, or *peritrichate*.

Fimbrine (pili)

Fimbrine are *short* and *thin hair-like filaments*, formed of protein called *pilin* (antigenic). They are responsible for the *attachment* of bacteria to *specific receptors* of the *human cell* (adherence). There are *special* types of pili called [sex pili], involved in the process of c*onjunction*.

Cell functions

Cell growth and metabolism

Between successive cell divisions, cells grow through the functioning of *cellular* metabolism. *Cell metabolism* is the process by which *individual* cells *process* nutrient molecules. *Metabolism* have *two* divisions: (1) *catabolism*, in which the cell breaks down complex molecules to produce energy and reducing power, and (2) *anabolism*, in which the cell uses energy that *reduces* power to construct complex molecules, along with performing other biological functions. Complex sugars *consumed* by the organisms can be *broken down* into a less chemically-complex sugar molecule called *glucose*. Once inside the cell, glucose is *broken down* to make *adenosine triphosphate* (ATP), a form of energy, via *two* different pathways.

The *first* pathway, **glucose**, which requires *no* oxygen, is referred to as *anaerobic metabolism*. Each reaction is designed to produce some hydrogen *ions* that can then be used to make *energy packets* (ATP). In eukaryotes, **glycolysis** is the *only* method used for *converting energy*.

The *second* pathway, called the **Krebs cycle**, or **acid cycle**, which occurs *inside* of the *mitochondria* and is capable of *generating* enough ATP to run all of the cell functions.

Creation of a new cell

Cell division involves a *single* cell (a mother cell) dividing into *two* daughter cells. This leads to *growth* in *multicellular* organisms (growth of tissue) and to *procreation* (vegetative reproduction) in *unicellular* organisms. *Prokaryotic* cells divide by *binary fission,* while *eukaryotic* cells usually undergo a process of *nuclear division*, called *mitosis*, followed by the *division* of the cell, called *cytokinesis*. A *diploid* cell may also undergo *meiosis* to produce *haploid* cells, usually *four*, which serve as *gametes* in multicellular organisms, *fusing* to form *new* diploid cells.

DNA *replication*, or the process of *duplicating* a cell's *genome*, is required every time a cell *divides*. Replication, like all new cellular activities, requires *specialized* proteins for carrying out the job.

Protein synthesis

Cells are capable of *synthesizing* new proteins, which of course, are *essential* for the *modulation* and *maintenance* of cellular activities. This process involves the formation of *new* protein molecules from *amino acid* building blocks, based on information *encoded* in DNA/RNA. Protein *synthesis* generally consists of *two* major steps: *transcription* and *translation*.

Transcription is the process where *genetic* information in DNA is used to produce a complimentary RNA strand. This RNA strand is then processed to give *messenger* RNA (mRNA), which is *free* to *migrate* through the *cell*. mRNA molecules *bind* to protein – RNA complexes called *ribosomes,* located in the *cytosol*, where they are *translated* into *polypeptide* sequences. The ribosome mediates the formation of a polypeptide based on the mRNA sequence. The mRNA sequence *directly* relates to the *polypeptide* sequence by *binding* to *translate* RNA (tRNA) adaptor molecules in *binding pockets* within the ribosome. The new polypeptide then *folds* into a functional *three-dimensional* protein molecule.

Cell movement or motility

Cells can *move* during many processes: such s *wound healing*, the *immune response* and *cancer metastasis*. For wound healing to occur, *white blood cells*, and cells that *ingest* bacteria *move* to the wound site to *kill* the *microorganisms* that cause infection. At the same time *fibroblasts* (connective tissue cells) move there to *remodel* damaged structures.

In the case of *tumor* development, cells from a *primary* tumor move away and spread to *other* parts of the *body*. Cell *motility* involves many *receptors, crosslinking, binding, adhesion, motor* and *other* proteins. The process is *divided* into *three steps* – (1) **protrusion** of the leading edge of the cell, (2) **adhesion** of the leading edge and (3) **deadhesion** at the cell *body* and *rear*, and *cytoskeletal* contraction to *pull* the cell *forward*. Each of these is driven by physical forces generated by *unique* segments of the *cytoskeleton*.

Origin of cells

The *origin* of **cells** has to do with the *origin* and *miracle* of **life** itself, **together** going *hand-and-hand*. The *birth* of the cell marked the *passage* from *prebiotic* chemistry to *biological life*.

Origin of the cell

The unit of *modern* organisms and *populations* of organisms is not yet *clear*, *natural selection* usually being proposed to work on the level of *genes, individual organisms, groups* of *organisms* and, of course – *species*. *None* of these models is *mutually- exclusive*, where *selection* may act on multiple levels *simultaneously*. In a *gene-centered* view of *evolution*, however, **life** is regarded in terms of **replications** – that is the DNA molecule in the organism. If *freely-floating* DNA molecules that code for *enzymes* were *not* enclosed in cells, the enzymes that benefit a given replication (by producing *nucleotides*, for example) would do so less efficiently, and would in fact benefit *competing* replications. If the *entire* DNA molecule of a replication is *enclosed* in a cell, then the *enzymes* encoded from the molecule will be kept close to the DNA molecule itself. The replication will *directly* benefit from its *enzymes*.

Biochemical, cell-like spheroids formed by *proteinoids* are observed by *heating* amino acids with *phosphoric* acid as a *catalyst*. They bear many of the *basic* features provided by *cell membranes*. Proteinoid-based *protocells* enclosing RNA molecules may have been the *first* cellular *life forms* on Earth. Some *amphiphiles* have the tendency to *spontaneously* form membranes in water. *Spherically* closed membrane contains *water* and is a *hypothetical precursor* to the *modern* cell membrane composed of *proteins* and *phospholipid bilayer* membranes.

Origin of eukaryotic cells seems to have evolved from a symbiotic community of **p**rokaryotic cells. It is almost certain DNA-bearing *organelles* like *mitichrondria* and the *chloroplasts* are what *remain* of ancient symbolic oxygen-breathing *protobacteria* and *cyanbacteria*, where the *rest* of the cell seems to be *derived* from an *ancestral archaean prokaryotic cell* – a theory termed **endosymbiotic theory**.

There is yet *still* considerable debate whether *organelles* like the *hydrogenosome* predicted the origin of *mitochondria*, or vice-versa.

Sex, as the stereotype choreography of *meiosis* and *syngamy* that *persists* in nearly *all extant eukaryotes*, may have *played* a role in the *transition* from *prokaryotes* to *eukaryotes*. An origin of *sex* as *viccination theory* suggests that the eukaryotic genome *accreted* from prokaryan *parasite* genomes in numerous rounds of *lateral* gene *transfer*.

Sex-as-syngamy (fusion sex) *arose* when *infected* hosts began *swapping* nuclearized genomes *coevolved*, vertically transmitted *symbionts* that conveyed *protection* against *infection* by more *virolent* symbionts.

More on DNA

DNA, or *deoxyribonucleic acid*, is the *hereditary* material in *humans* and almost all other *organisms*. Nearly every *cell* in a person's *body* has the *same* DNA. Most DNA is *located* in the cell *nucleus* (where it is called *nuclear DNA*), and a small amount of DNA can also be found in the *mitrochondria* (where it is called *mitrochondria DNA* or mtDNA).

The *information* in DNA is *stored* as a *code* made up of *four* chemical *bases*: *adenine* (A), *guanine* (G), *cytosine* (C), and *thymine* (T). *Human DNA* consists of about **3 million basis**, and *more* than **99 percent** of these *bases* are the **same** in **all people**. The *order*, or *sequence*, of these *bases* determine the *information* available for *building* and m*aintaining* an *organism,* is similar to the way in which *letters* of an *alphabet* appear in a certain *order* to form *words* and *sentences*.

DNA bases *pair up* with *each other*, A *with* T *and* C *with* G, to form *units* called **base pairs**. Each *base* is also attached to a **sugar** *molecule* and a **phosphate** *molecule*.

Together, a base, sugar, and phosphate are called nucleotide. Nucleotides are arranged in a *base, sugar*, and *phosphate* are called *nucleotide*. **Nucleotides** are *arranged* in **two** long s**trands** that form a **spiral** *called* a **double helix**. The structure of the *double helix* is somewhat like a **ladder**, with the **base pairs** *forming* the *ladder's* **rungs** and the **sugar** and **phosphate** *molecules* forming the **vertical sidepieces** of the **ladder**.

An important *property* of DNA is that it can **replicate**, or make *copies* of **itself**. Each s*trand* of DNA in the *double helix* can serve as a *pattern* for *duplicating* the *sequence* of b*ases*. This is **critical** when *cells* **divide** because *each* **new cell** needs to *have* an **exact copy** of the DNA *present* in the **old cell**.

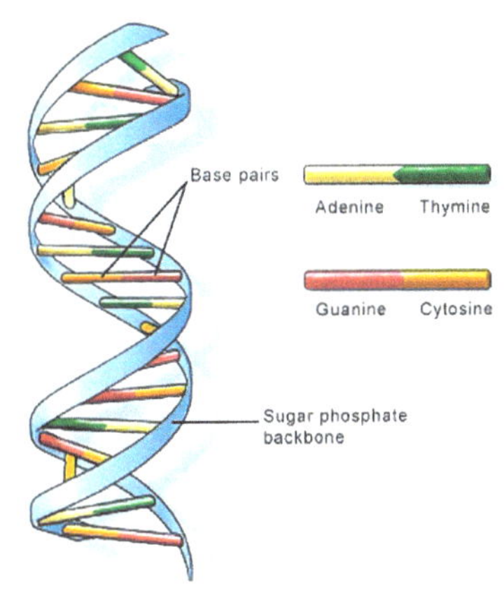

DNA Structure

DNA is sometimes called "**the blueprint of life**" because it contains the code, or i*nstructions* for *building* an organism and insuring that an organism functions *correctly*. just as a builder uses a *blueprint* to b*uild* a house, DNA is used as the b*lueprint*, or *plan*, for the *entire* organism. The *shape* of the DNA molecule is a *double-helix* (like a t*wisted ladder*) The sides are *composed* of alternating *sugars* (deoxyri-bose) and *phosphates*. The *rungs* of the ladder are com- posed of *nucleotides*. Nucleo- tides – also called *Bases*: *Adenine, Thymine, Cytosine*, or A.T.G.C. Nucleotides pair in a specific way – called the Base-Pair-Pair Rule Adenine pairs to Thymine Memory helper – think could "A T Granite City" which is where you live. The rungs of the ladder can occur in any order – as long as the base pair is followed All in all there are billions of bases (molecules) in cells, coding for all things cells need.

More DNA Replication

Replication is the *process* where DNA *makes* a *copy* of *itself.* **Why** does DNA *need* to **copy**? Simple: *Cells* divide for an *organism* to *grow* or *reproduce*. Every **new cell** *needs* a *copy* of the DNA, or *new instructions* to know *how* to *be* a *cell*. DNA *replicates* right before *cell* **divides**.

DNA *replication* is **semi-conservative**. That means that when it *makes* a *copy*, one half of the *old strand* is always *kept* in the *new strand*. This helps *reduce* the *number* of copy *errors*.

More on RNA

DNA *remains* in the *nucleus*, but in order for it to get its instructions *translated* into p*rotein*, it must *send* its *message* to the *ribosomes*, where *proteins* are *made*. The c*hemical* used to *carry* this *message* is Messenger RNA. mRNA has the job of taking the message from the DNA to the nucleus of the ribosome.

More on Genes

Most **genes** *contain* the *information* needed to *make* functional *molecules* called *proteins* (a few genes *produce* other *molecules* that *help* the cell *assemble* proteins.) The *journey* from *gene* to *protein* is *complex* and *tightly* controlled *within* each *cell*. It *consists* of two m*ajor steps*: **transcription** and **translation**. Together, *transcription* and *translation* are k*nown* as **gene expression**. During the *process* of **translation**, the *information* stored in a *gene's* DNA is t*ransferred* to a similar *molecule* called RNA (ribonucleic acid) in the cell *nucleus*. Both RNA and DNA are made up of a *chain* of *nucleotide bases*, but they have slightly d*ifferent properties*. The *type* of RNA that *contains* the *information* for *making* a *protein* is called a *messenger* RNA (mRNA) because it *carries* the *information*, or *message*, from the DNA *out* of the *nucleus* into the *cytoplasm*.

Translation, the *second step* in getting from a *gene* to a *protein*, takes place in the *cytoplasm*. The mRNA *interacts* with the specialized *complex* called a *ribosome*, which "reads" the *sequence* of mRNA *bases*. Each *sequence* of these *bases*, called a *codon* usually *codes* for *one* particular *amino acid* – *amino acids* are the *building blocks* of *proteins*. A type of RNA called *transfer* RNA (tRNA) *assembles* the *protein*, *one* amino acid at a time. *Protein* assembly *continues* until the *ribosome* encounters a "stop" *coden:* a sequence of *three bases* that does *not code* for an *amino acid*.

The *flow* of *information* from DNA to RNA to *proteins* is one of the **fundamental principles** of **molecular biology**. It is so *important* that it is sometimes *called* the "**central dogma**."

The Difference Between Genes and Alleles

A **gene** is an *information unit* that *describes* how to *construct* a particular *type* of **molecule** – Lets leave *aside* for the moment how the *information* is "written" and *who/what* does the *construction*. By *themselves* genes *can't* do anything *other* than provide *information* on to how to *make* something that *does* something. In other words, a *gene* is simply a *set* of *instructions* that *describes* how to *build* a *specific* type of *molecule* that will *perform* some *concrete* active *job* in the *cell*. In general, you can think of the *molecule* described by a *gene* as a *kind* of a machine *designed* for a specific *job*, usually some *biochemical reaction*.

For any *given* **gene**, *plants* and *animals* usually carry *two copies*. A **key thing** to **know** is that in any *specific gene*, there can be *different versions* (also *called* **alleles**).

Otherwise, the *two* copies of a *gene* in an *organism* are *different versions* of the same *gene*, or *sometimes* the *same version*. Here's a crude *analogy* to illustrate:

A friend of mine ***baked*** a ***cake*** – the ***recipe*** called for ***one*** cup of ***oil***, but didn't ***specify*** w**hich** kind of ***oil***. ***Vegetable oil? Peanut oil? Whale oil? Pennzoil?*** All of these were o***ils***, but they would no doubt produce cakes that taste ***quite different***. Now imagine the situation if instead of ***one*** cup of oil, the recipe called for ***two*** cups – and you could m***ix*** different ***types*** of ***oil*** in your ***cake***. Here are some ***possibilities***:

vegetable oil vegetable oil

vegetable oil peanut oil

vegetable oil whale oil

vegetable oil Pennzoil

peanut oil peanut oil

peanut oil peanut oil

peanut oil Pennzoil

whale oil whale oil

whale oil Pennzoil

Pennzoil Pennzoil

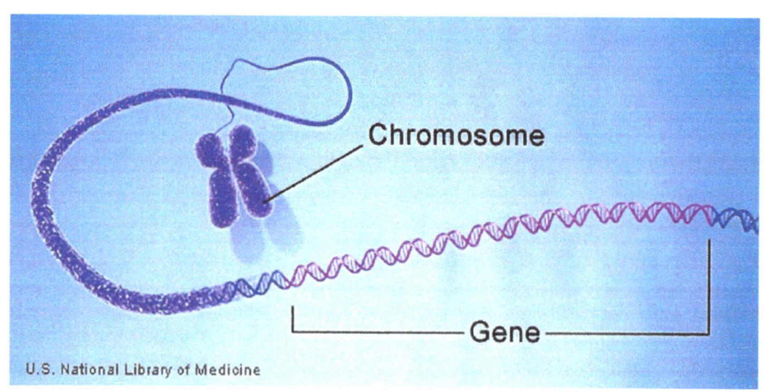

DNA is made up of genes. Each chromosome contains many genes

As you can no doubt *see*, you'd get some *weird cakes* with *some* of these combinations. This baking *example* will hopefully *illustrate* something to you about *genetics* – each *different* kind of *cake* oil being an "allele" *of* oil. For each gene, h*umans* and most *orchids* carry *two copies* (there are exceptions). One *copy* comes from the *mother*, and the other from the *father*. Now, each *copy* can be one of p*otentially* many *alleles* that *exist* in the *general population*.

Let's *assume* that we are *talking* about just one gene, call it gene G. Assume also that t*welve* other alleles of this *gene* exist out of this *population*. Although everyone (orchid and human) in the *population* has *two copies* of the *gene*, a *particular* individual may c*arry* any *two* of the *twelve* possible *alleles* – in fact, an *individual* may carry *two copies* of the *same allele*, an exceedingly *common occurrence*.

In the above *illustration*, *each* parent *carries* two different *alleles*, and these parents' a*lleles* are *different* from each other. Hence, during *mating*, these parents could only c*ontribute* to their offspring *one* of the *four* combinations of alleles *shown*.

OK. Now that we understand *difference* between *genes* and *alleles* more *clearly*, let's go *back* to how *alleles* of *genes* (i.e., different versions of genes) *affect* **traits**. A trait is *basically* what you see. It is the *physical* expression of the *genes* themselves. *Brown hair, blue eyes, yellow flowers, long petals* – all of these traits *derive* from some *combination* of *alleles*. The *term* for this *physical* expression is phenotype, and the *corresponding* term, genotype *describes* the *genes* that *produce* the *phenotype*.

In his *plant breeding* experiments, **Mendel** studied pea plant *phenotypes* that p*resented* themselves *clearly*. The *pea seeds*, for example, were either *smooth* or w*rinkled*, the flowers were *white* or *violet*, the pods *full* or *constricted*. His *selection* of clearly discernible, unblended *traits* proved critical to Mendel's *conclusions* of how *traits* p*assed* from one *generation* to *another*.

More on DNA, RNA, and Proteins

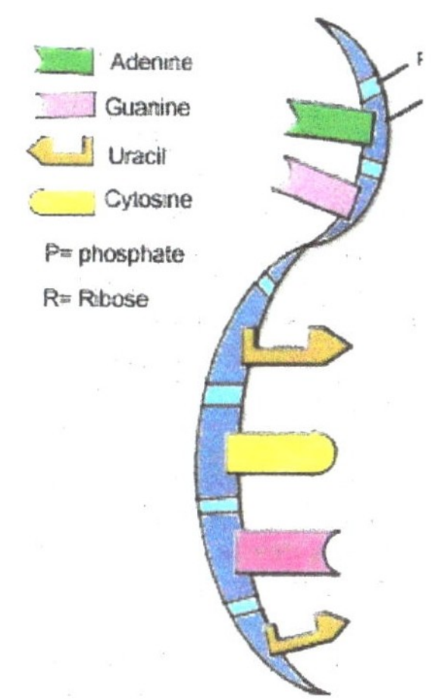

What were the first molecules to form life?
Since **we** weren't *around* to *observe* the *beginnings* of **life**, we must use *clues* found in the *structures* and *functions* of today's biological *molecules* to *answer* this *question*.

Every **living thing** is **descended** from its **ancestors**. *Evolution* results when **changes** o*ccur* along the *way*.

At the **molecular** *level*, the *same* applies. Each *genetic* information *molecule* must be c*opied* and *passed* on to future *generations*. These molecules reveal *clues* about their *life* on *Earth*. The *behavior* of today's biological molecules reveals *clues* about potential r*oles* in the origin of *life*.

What do these molecules do today?

RNA

What is it and what does it do?

RNA is the *polymer* made up of *four* repeating *subunits* called **nucleotides**. Each n*ucleotide* contains a **sugar**, a *phosphate* **group** and a *nitrogenous* **base**. In RNA, that s*ugar* is **ribose** (In DNA, it's **deoxyribose**). The *nucleotides* that *comprise* each RNA s*trand* occur in a specific *sequence*. In *cells*, the *sequence* of an RNA strand is *specified* by a *corresponding* sequence in the *cell's* DNA. RNA bases can pair with one another if they are complementary: U (uracil) pairs with A (adenine) C (cytosine) pairs wit G (guanine) Even so, RNA generally occurs as a **single strand**. A single strand *molecule* may *fold* in such a way that complemenary bases are brought together to pair along short stretches. The sequence of RNA bases along short stretches. The sequence of RNA bases along the strand determines how it will fold. In this way, the sequence of bases determines the strand's 3- dimensional structure. What an RNA molecule can do depends in part on its *three- dimensional shape*.

RNA = ribonocleicacid
RNA is similar to
DNA except:

1. has one strand instead of two strands
2) has uracil instead of thymine
3) has ribose instead of deoxyribose

1. RNA can store genetic information

A *genome* is the *catalog* of *genetic* material in a *living* system. For most *organisms*, the *genome* is *made* of DNA. But, in some *viruses*, the *genome* is *made* of RNA, **not** DNA. Human *immunodeficiency* virus **type 1** (HIV1), the virus that *causes* AIDS, is an RNA **virus**.

2. A genome can be a temporary copy of genetic information

Messenger RNA (mRNA) is a short-livid intermediary *molecule* that *contains* the i*nformation* for *making* a specific *protein*. This *information* is first *copied* from a *cell's* DNA *genome*, and then the mRNA is *taken* to the cell's protein synthesis machinery, *called* the **ribosome**. By "reading" the information *encoded* in the mRNA, the *ribosome* knows which specific **protein** to make.

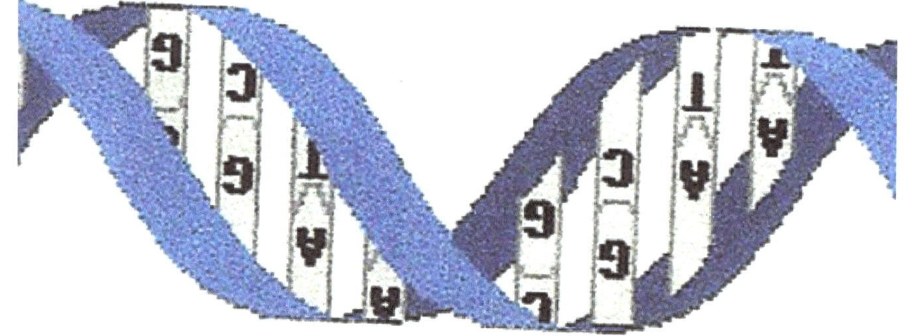

Transcription – RNA is made from DNA
Proteins are made from the message on DNA
DNA to mRNA to RNA

3. RNA can be carrier molecule

One role of RNA is to *transport* other *molecules* needed for *specific* biological r*eactions*. Transfer RNA (tRNA) plays this *role* in protein *synthesis*. There are **20** d*ifferent* tRNA *molecules*, each of which is *programmed* to *attach* specifically to one of the 20 **amino acids** found in *proteins*. tRNAs brings these *amino acids* to the r*ibosome*, where they are *incorporated* into the *growing* protein *chain* according to s*pecifications laid out* in the mRNA.

4. RNA can be a working part of cellular machinery
The **ribosome** is a *large* structure *composed* of more than **50** *different proteins* and several *ribosomal* tRNA *molecules.* rRNAs are *crucial* to protein *synthesis* because they *form* the structural *core* of the *ribosome.*

5. RNA can be a catalyst
Catalysts drive *reactions* between *molecules.* They *reduce* the *amount* of *energy* required for the *reactions* to *occur*, and they make sure the *reaction* is *specific* and *accurate.* Protein *catalysts* are known as **enzymes**, and RNA can also *behave* as a *catalyst.* In fact, the *ribosome* relies on the *catalyst* functions of rRNA to *assemble* proteins.

RNA *catalysts* are called **ribozymes**. *Ribozymes* can "*cut* and *paste*" *links* between *nucleotides* in a *strand, a* process *known* as *splicing*. Some mRNA molecules m*ust splice* themselves before they can be *accurately* "read" by the *ribosome* during p*rotein synthesis.*

Protein synthesis could **not** take place **without** these *types* of RNA, the *Messenger*; mRNA, the *Carrier*, tRNA, the *Machinery*, rRNA.

DNA
What it and what it does, DNA is the polymer made up of *four* repeating subunits called n*ucleotides.* Each molecule contains a sugar, a *phosphate* group and a nitrogenous. In DNA, that *sugar* is d*eoxyribose* (In DNA, it's *ribose*).

DNA can *occur* as a **single** or **double-stranded** *molecule.* The double-stranded *form*, which **twists** into a **helix**, is a very *stable* structure. Complementary *bases* along the t*wo strands* **pair** in a **specific** way: A (adenine pairs with T (thymine) B (cytosine) pairs with G (quanine) DNA can store genetic information. **Nucleotides** occur in a *specific* s*equence* within each *strand* of a DNA *nucleotide.* A relatively short *stretch* of DNA that e**ncodes** a *protein* or DNA *molecule* is called a **gene**. DNA *serves* as a *stable repository* of genetic *information* of organisms in all *five kingdoms.*

Proteins
Proteins are *polymers* made of **amino acids**. There are **20** *different* amino acids commonly *found* in *proteins.* To *make* a *protein*, the *ribosome* strings amino acids t*ogether* into a *chain.* This *chain*, called a *polypeptide*, folds into a specific *three- dimensional* shape.

The **first shape** of a **protein** determines its *function.* Many *enzymes*, for example, are g*lobular*, while structural *proteins* often take a *long, fibrous* shape.

1. Proteins are products of genetic information
Amino acids are h*eld together* in a protein *chain* by strong bonds. The specific order of amino acids in the chain is specified by information encoded in genes, which are relatively short stretches of DNA within the genome of a living system. An intermediate copy of the messenger RNA (mRNA directs synthesis of a protein by listing the order in which amino acids will be added to the protein chain.

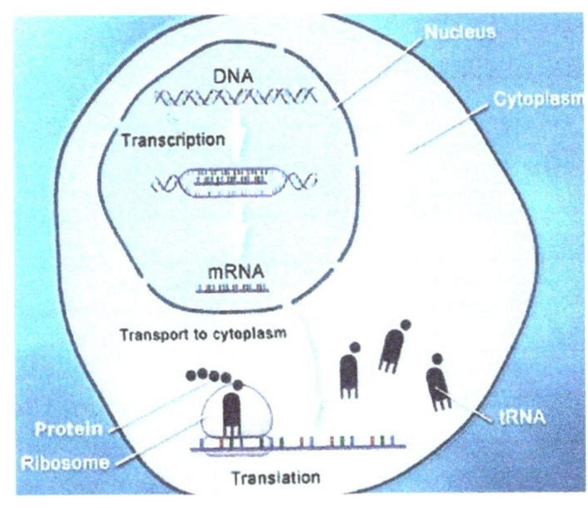

Through the processes of transportation and translation, information from genes is used to make proteins

2. Proteins can be catalysts
Catalysts drive *reactions* between *molecules.* They *reduce* the amount of *energy* r*equired* for the *reaction* to *occur*, and they make *sure* the *reaction* is *specific* and a*ccurate*. Protein *catalysts* are known as *enzymes*. Enzymes make the huge *variety* of biochemical *reaction* in the cell *possible* by *keeping* the *processes* very *specific.*

Enzymes, in addition, help *reactions* take place in reasonable *time* at the t*emperatures* found in *living systems*.

What does this tell us about the first molecules of life?

The central dogma of biology tells us:

1. Information stored in DNA is eeded to replicate DNA

2. Information stored in DNA is needed to make proteins.

3. RNA can serve as:
   a. The **Messenger** – mRNA is an *intermediate* copy of *genetic* information that g*uides* protein *synsthesis*.
   b. The **Carrier** – tRNA carries *amino acids* to the *ribosome*.
   c. The **Machinery** – rRNA is an *essential component* of the *ribosome*.

4. Protein catalysts (enzymes) are *needed* to interpret *genetic* information, but they are also the *product* of this *interprettion*.

Which molecules came first?

Based on the *modern roles* and *behaviors* of DNA RNA and *proteins*, which *molecule* would you *elect* as the most likely *candidate* for the *molecular* origin of *life*? Why?

Life science *researchers* have been *pondering* this *question* for *years*. What do they t*hink*? Explore this in *depth* in the *monograph* Bringing RNA Into View: *RNA and its Roles in Biology*.

More on Micrchondria DNA

Although mot DNA is *packaged* in *chromosomes* within the *nucleus*, **mitrochrondria a**lso have a small *amount* of their *own* DNA. This genetic material is known as as *mitochondrial DNA* or **mtDNA**.

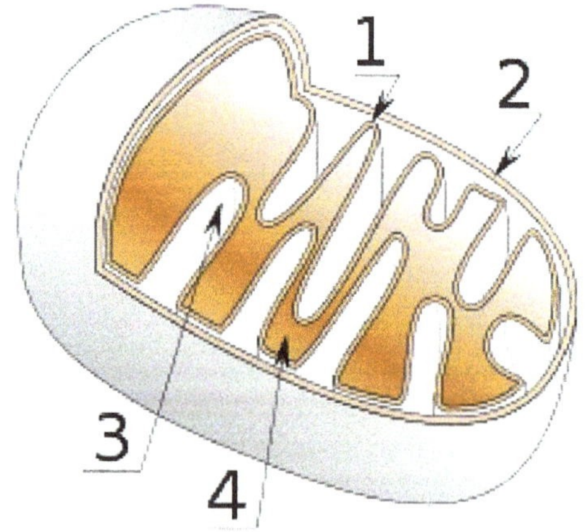

Mitochondria Structure

Mitochondria are structures within *cells* that *convert energy* from *food* into a form that cells can use. Each *cell* contains *hundreds* to *thousands* of *mitochondria*, which are l*ocated* in the *fluid* that *surrounds* the *nucleus* (the cytoplasm). *Mitochondria* produce *energy* through a *process* called *oxidative phosphorylation*. This *process* uses *oxygen* and simple *sugars* to create *aderosine triphosphate* (ATP), the *cell's* main *energy* source. A set of enz*yme complexes*, designed as *complexes 1-V*, carrry out *oxidationphosp-* horylation within mitochondria. In *addition* to energy production, *mitochondria* play a role in *several* other c*ellular* activities. *Mitochondria*, for example, help r*egulate* the *self-destruction* of cells (apoptosis). They are also *necessary* for the *production* of substances such as cholestrol and a heme (a component of *hemo- globin*, the molecule that carries *oxygen* in the *blood*). *Mitichondria* DNA contains **37 genes**, all of which are *essential* for making *enzymes* involved in *oxidative phosphor- ylation*. The remaining *genes* provide *instructions* for making *molecules* called t*ransfer* RNAs (tRNAs) and *ribosomal* RNAs (rRNAs), which are **chemical cousins** of DNA. These types of RNA help *assemble* protein *building blocks* (amino acids) into functioning *proteins. A* **gene** is the *basic physical* and *functional unit* of *heredity*. Genes, which are made up of DNA, act as *instructions* to make *molecules* called *proteins*. In h*umans*, genes vary in *size*, from a **few hundred** DNA *bases* to more than **2 million bases**. The *Human genome Project* has estimated that *humans* have between **20,000** and **25,000 genes** in each **cell**. Every *person* has **two copies** of **each gene**, one *inherited* from e**ach parent**. Most *genes* are the *same* in *all people*, but a small *number* of *genes* are *slightly* different *between* people. **Alleles** are *forms* of these same *gene* with small *differences* in their *sequence* of DNA *bases*. These small *differences* contribute to each person's *unique physical features*.

Let's assume we're talking about just one gene, and call it gene G. Assume also that twelve different alleles of this gene exist out in the population. Although everyone (orchard or human) in the population has two copies of the gene, and a particular individual may carry any two of the twelve possible alleles – In fact, an individual may carry two copies of the same allele, which is an exceedingly common occurrence.

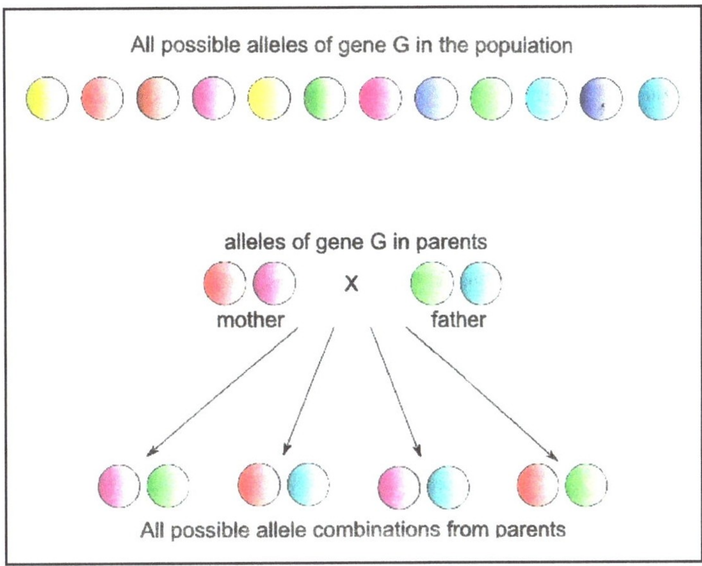

In the above illustration, each parent carries two different alleles, and all four of these parents are different from each other. Hence, during a mating, these parents could only contribute to their offspring one of the four combinations of alleles shown.

OK, now that we understand more clearly the difference between genes and alleles, let's go back to how alleles of genes (i.e., different versions of genes) affect traits.

A trait is basically what you see. It is the physical expression of the genes themselves. Brown hair, blue eyes, yellow flowers. Long petals – all of these traits derive from some combination of alleles. The term for this physical expression is phenotype. The corresponding term, genotype describes the genes that produce the phenotype.

In his plant breeding experiments, Mendel studied pea plant phenotypes that presented themselves clearly. The pea seeds, for example, were either smooth or wrinkled, the flowers were white or violet, the pods full or constricted. His selection of clearly discernable, *unblended* truly proved *critical* to Mendel's conclusions of how traits are *passed* from *one* generation to *another*.

More on Chromosomes

In the **nucleus** of *each cell*, the DNA *molecule* is *packaged* into *thread-like structures* called **chromosomes**. Each *chromosome* is made up of DNA *tightly coiled* many times around **proteins** called **histones** that *support* the *structure*.

Chromosomes are **not** *visible* in the cell's **nucleus** – not even under a **microscope** – when the *cell* is **not** *dividing*. The DNA that make up *chromosomes*, however, *becomes* more *tightly packed* during **cell division** and is then **visible** under a **microscope**. *Most* of what researchers *know* about *chromosomes* was *learned* by observing *chromosomes* during **cell division**.

Each *chromosome* has a *constriction* point called the **centromore**, which *divides* the c*hromosome* into *sections*, or "arms." The *short arm* of the *chromosome* is labeled the "**p arm**." The *long arm* of the *chromosome* is labeled the "**q arm**." The *location* of the c*entromore* on each *chromosome* gives the *chromosome* its characteristic *shape*, and can be used to help *describe* the *location* of *specific genes*.

The shape of the

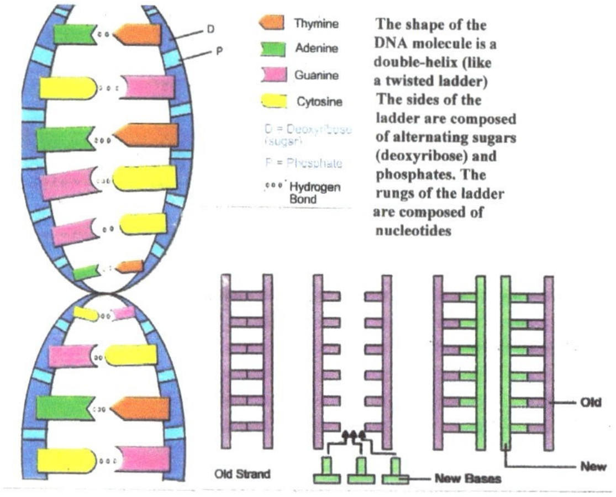

Nucleotides (also called bases)

Adenine Thymine, Guanine or A.T.G. C

Nucleotides pair in a specific way – called Base-Pair-Rule

Adernine pairs to Thymine

Guinine pairs to Cytosine

Memory helper – think ("AT Granite City"), which is where you live

DNA molecule is a double-helix (like a twisted ladder)
The sides of the ladder are composed of alternating sugars
(deoxyribose) and phosphates. The rungs of the ladder are composed of nucleotides.

Nucleotides (also called bases)
Adenine Thymine, Guanine or A.T.G. C
Nucleotides pair in a specific way – called Base-Pair-Rule
Adernine pairs to Thymine
Guinine pairs to Cytosine
Memory helper – think ("AT Granite City"), which is where you live

In *humans*, each cell normally *contains* **23 pairs** of *chromosomes*, for a total of **46**. **Twenty-two** of these **pairs**, called **autosomes**, look the *same* in both *males* and f*emales*. The **23rd pair**, the **sex chromosomes**, differ between *males* and *females*. Females have *two copies* of the **X chromosome**, while males have **one X** and **one Y** c*hromosome*.

The **22 autosomes** are *numbered* by **size**. The *other* two *chromosomes*, **X** and **Y**, are the **sex chromosomes**. This picture of the **human chromosome** *lined up* in **pairs** is c*alled* a **karyotype**.

Chromosomes and Heredity

Why do *children* so often *resemble* their parents? Why do some *brothers* and *sisters* share *similar traits*, while others are *very different*? To a large degree, it's a function of the *genes* – which are the *basic units* of *heredity* – they have in *common*. How does this happen? Let's find out a little bit about what genes *are* and *how* we *inherit* them.

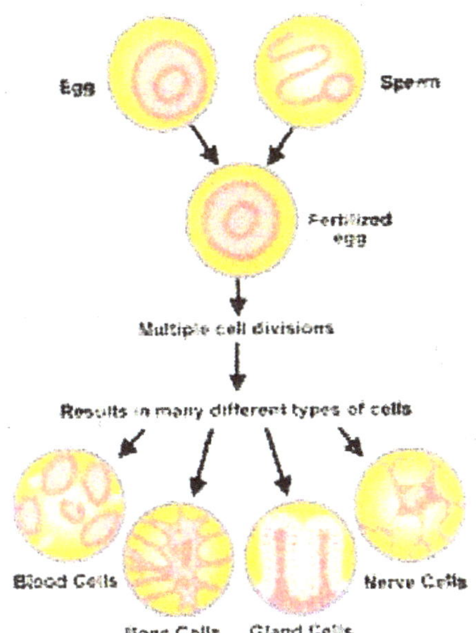

Your **body** is made up of **billions** of **cells**. In some ways, your *cells* can be very *different* from each other. They can, for example, specialize in a particular function, such as *carrying oxygen* (red blood cells), *absorbing food* (intestinal cells), or *sensing light* (cells in your eyes).

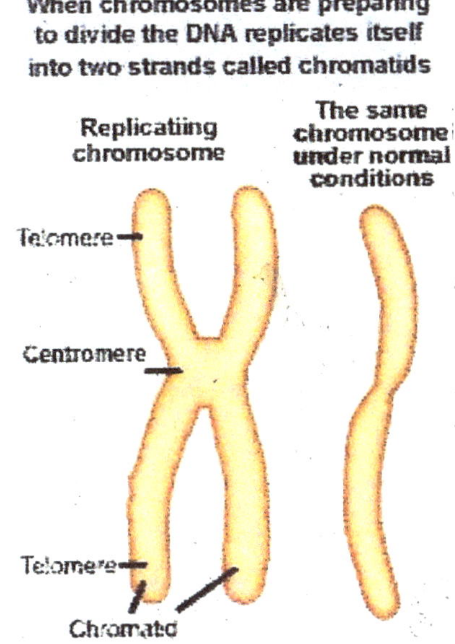

In other ways, your cells have a *lot* in *common*. At the *center* of almost all your *cells*, for instance, is a *ball-shaped* structure called the n**ucleus**, inside of which are **46** *thread-like structures* called **chromosomes**, these chromosomes contain the estimated **35,000 genes** that make us who we are.

To *understand* how we end up with a *given* set of *genes*, we need to learn more about **chromosomes, DNA,** packaged together with **proteins** and other kinds of **molecules**. Each chromosome has a **centromere**, which plays an important *role* during cell *division*, and also *divides* each *chromosome* into a long arm.

Scientists can tell *different* chromosomes *apart* based on their *size*, the relative *lengths* of their *arms*, distinctive *patterns*, and other *characteristics*.

Humans have two types of chromosomes: **sex chromosomes** and **autosomes**. *Two sex chromosomes* determine the **sex** of an individual, and they are called the **X** *chromosome*, and the **Y** chromosome.

If you are **female**, you have two **Xs,** and if you are **male**, you have one **X** (although there are *genetic conditions* in which this *varies*). The *autosomes* comprise the other **22** c*hromosomes*. The *longest* of the *autosomes* is referred to as *chromosome* **1**, the next largest as *chromosome* **2**, and so on, down to the *smallest* autosome, chromosomes **21** and **22**.

Each cell *nucleus* contains *two copies* of *each autosome* (44 chromosomes), plus *two* s*ex* chromosomes (either **two Xs** or **an X** and **a Y**) for a total of **46**. With *few* exceptions, the *chromosomes* and *genes* found within any two cells of *your body* will be *identical*.

The *mystery* as to why you *resemble* your *family members* is *solved* by discovering how you *inherited* your *chromosomes* from your *parents*.

Each *chromosome* within our *body* (except the chromosomes within cells that develop into *sperm* or *egg*) is *created* by making a *copy* of a *previously* existing c*hromosome*. This occurs during the process called **mitosis**, during which cells *divide* for g*rowth* or *repair*. Before each *division*, the cell makes an *identical copy* of each chromosome and during *mitosis*, each of the *two new cells* receives a complete set of **46** c*hromosomes*.

Each *new cell* has the same set of *chromosomes* and the same *genetic information* as to the "parent" cell. This *explains* why almost *every cell* in *your body* has the same *genetic information.*

Cell Division

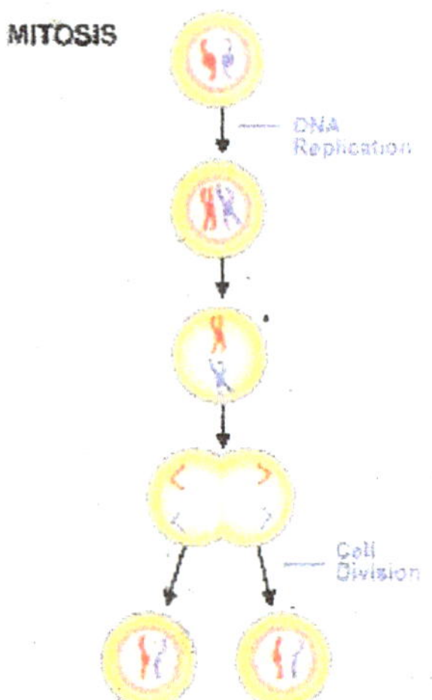

A slightly *different* process takes place during the production of *eggs* and *sperm cells*. Where an *egg* and a *sperm* unite at *fertilization*, their nuclei *unite* to *form* the *nucleus* of a *human zygote*. If the *sperm* and *egg* carried **46** *chromosomes*, like the rest of the body's *cells*, then the *rest* of the body's *cells* would have **92**, which would be incompatible with life.

To *prevent this*, a special type of *cell division*, called m**eiosis**, takes place. The process of *meiosis* begins with a *single cell* containing **46** *chromosomes* and results in *four* reproductive cells (*sperm* and *egg*), each of which carries **3** *chromosomes*.

An *important feature* of these *four cells* is that the *combination* of *genes* they *carry* on their **23** *chromosomes* is a *unique mix* of the *genes* present in the original *single cell*.

You *resemble* your *parents* because *half* of the *instructions* – genes – for making you came from your *father* and *half* from your *mother*. Similarly, your *brother* or *sister* also received *half* of their genetic *instructions* from each *parent*, but the *set* they received is somewhat *different* from the *set* you received. That's why they may *resemble* you, but they are not *identical* to you. *Identical twins* receive *exactly* the same *combination* of genes and *chromosomes*.

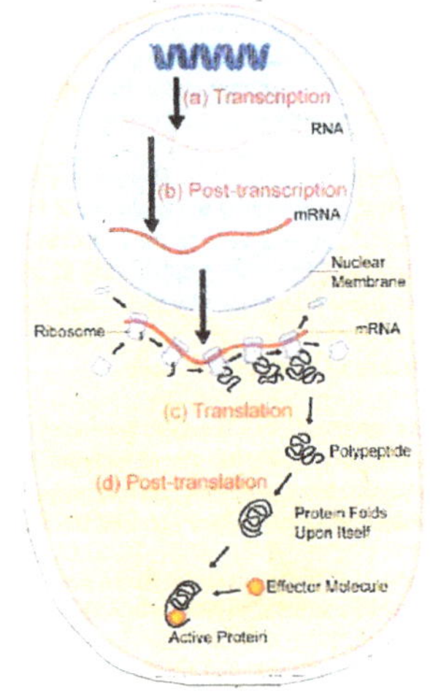

DNA and Proteins

As discussed in the Basic Genetics section, **DNA** is a *large molecule* packaged in *chromosomes* in the **nucleus** of *cells*. The DNA *molecule* contains genes that *direct* the *production* of *proteins*.

Proteins are *molecules* that play a *critical role* in the *structure* of your body's *cells, tissues*, and *organs*. Every *protein* is made up of *building blocks* called *amino acids*. The *code* that is *carried* by DNA *determines* which *amino acid* will come *together* in what *order* to *form* a given *protein*. Genes *act*, or "*express*," themselves by *dictating* the *order* of *amino acids* used to make *proteins*. The *proteins* made by some *genes* are needed by *all cells*, but different *sets* of *genes* may be switched *on* or *off* in *different* collections of *proteins* being made and results in *different structures, appearances* and *functions*. In addition to *determining* what *proteins* are made, the DNA in a cell also *controls* how much of a *protein* will be *made* and under what *circumstances*.

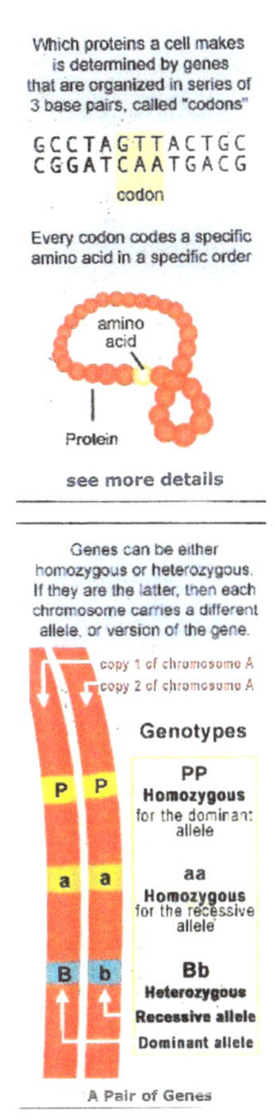

Heredity

If you needed several *family trees* and *traced* the i*nheritance* of a *given trait* in the *families*, you would find that *unique patterns exist*. Several *factors* are involved in determining *patterns of inheritance*, including where the *trait-causing gene* is *located* (on the autosomes or sex chromosomes) and whether *one* or *two* copies are necessary for a *given trait* to be expressed. *Genes* may *exist* in *more* than *one form*, each of which is called an **allele**; the most *common form* of a *gene* is called a "wild type." No matter how many forms (or alleles) a gene has, each person *inherits* only *two* of them – one from the *mother* and one from the f*ather*.

Genotype (the *pair of alleles* a person has at a *specific location* in the *genome*) affects **phenotype** (the effect of the *allele*, such as *eye color*; in the case of *medication*, how a person *reacts* to a *drug*).

Gene *variation* (alleles) may *change* the *gene* so that it *codes* for a *protein* that *works* just as *well* or *better* than the protein *coded* for by the "wild type." *Variant alleles*, however, can also *change* a *protein* so that it no longer *works as well* or does *not work at all*.

To *illustrate*, let's consider *genes* that *code* for the *proteins* (called **enzymes**) that break down *unnaturally occurring chemicals* in our body (*air pollutants, poisons*, or m*edicines*, for example) so they can be *safely excreted*.

Suppose that in a *certain population*, such as everyone living in the *United States*, such a *gene* is present in *two forms*, called "**A**" and "**B**." We would say *this gene* has *two alleles*, Allele "A" codes for an enzyme that breaks down Drug X quite efficiently, and it is most *common* in the *population* (wild type).

Allele "**A**" resulted from a **mutation** (*change* in the *order* of DNA *bases*) in allele "**A**" at some point in *evolution*, and the *enzyme* it produces doesn't *break down* **Drug X** at all.

While *people* who carry *two copies* of the "**A**" allele can take Drug **X** safely, *people* who carry *two copies* of the "**a**" allele, on the other hand, will *not* be able to *break down* Drug **X** effectively, so *large amounts* of the *drug* will *stay* in their *bodies* for a *long time*, which might lead to *serious side-effects*. People who carry one copy of "**A**" and one of "**a**" may have an *intermediate response* (depending on *how* the *enzyme* is *affected*). This example *illustrates* how *important* a person's *genes* can be when it come to being treated s*afely* and *successfully* with a *medicine*.

Inheritance Patterns

Each *gene* is present in every *chromosome* (generally), so a *pair* of *chromosomes* contains *two copies* of the same gene. The two copies may be *identical* or *different*. In cases where there are *variations* of a *gene*, one of the *alleles* can take *precedence* over (or override) the *other*. In the *classic case* observed by **Mendel**, the result of *crossing* a *white-flowered* pea with a *purple-flowered* pea was a p*urple* in a *purple-flowered offspring* – not a *pale purple*, which would suggest *mixing* or *blending* of the t*wo genes*. Conversely, the gene for *purple* color was d*ominant* over the gene for *white* color, which is called **recessive**.

A person who has *two identical alleles* for a *gene* is said to be **homogeneous** for *that* gene. A person with t*wo different alleles*, on the other hand, is said to be h**eterogeneous**. The *dominant* allele is usually r*epresented* by a *capital letter* and, the *recessive* allele by a l*ower case* letter (i.e., **A**, **a**).

A *person* (or *pea plant*) with two *dominant alleles* is called *homogeneous* dominant for that *gene* – and that pea plant would have *purple* flowers. With one *dominant* and one *recessive* allele, the person would be *heterogeneous* – and the *pea* would have *purple* flowers. If two *recessive* alleles are present, the person is *purple* flowers. If the *person* is h*omogeneous recessive* – and only in this case, with both alleles coding for *white* flowers, would the pea plant have *white* flowers.

1. *Autosomal dominant inheritance* is caused by a *mutation* in a *gene* located on an a*utosomal chromosome* and *occurs* when one *autosomal* allele *masks* the e*xpression* of another *allele.* A dominant *gene* from just *one* parent will result in the *phenotype,* which may be *eye* or *hair color* or may be a *serious condition* such as *Huntington's Disease.*

2. *Autosomal recessive inheritance* is caused also by a *mutation* in a gene *located* on an autosome; but in this case, *two copies* of the *recessive gene* are needed for the t*rait* to be *expressed.* **Cystic fibrosis** and diseases affecting *metabolism* (such as p**henylketonuria/PKU**) are autosomal recessive conditions.

3. *X-linked recessive inheritance* also is *caused* by a gene on the **X** *chromosome* rather than on an *autosome.* Females have two **X** *chromosomes,* so they need to inherit *two copies* of the *allele* to *express* the *trait*; if they have just *one copy,* they are a *carrier* of the *trait,* but do *not* exhibit *it* (although they may have *mild symptoms* in *some cases*). Males are *affected* by **X**-linked *recessive traits* because they have only **X** chromosomes, so they need only *one variant gene* to express the t*rait.* **Hemophilia** is an example of an **X**-linked *recessive disease* – it affects men exclusively.

4. *Complex inheritance* involves the *additive effect* of many genes *interacting* with e*ach other* and with the *environment.* Common diseases such as *heart disease, obesity, osteoarthritis* and *asthma* are *not inherited* according to Mendel's p*atterns,* but *result* from an *interplay* of *environmental* factors (such as *exercise, smoking,* and *exposure* to *pollutants*).

What are chromosomes and why do we need them?

Chromosomes are compact *spools of DNA*. If you were to *stretch out all* the DNA from *one* of your *cells*, it would be over 3 feet (1 meter) l*ong* from *end to end!* You can think of *chromosomes* as "DNA p*ackages*" that enable all this DNA to *fit* in the *nucleus* of *each cell*. Normally, we have 46 of these *packages* in *each cell*; we receive 23 from our *mother* and 23 from our *father*.

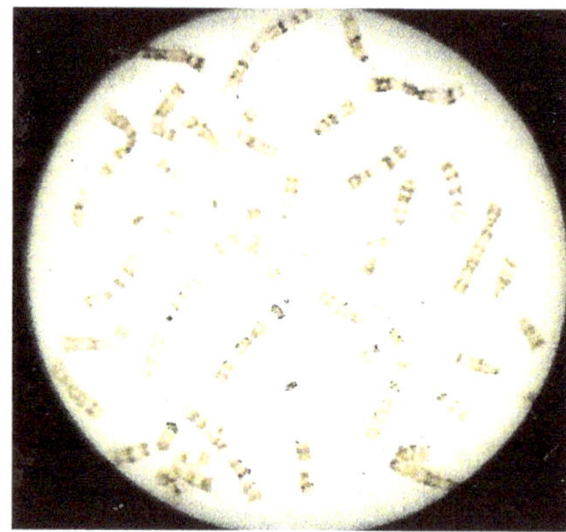

Here is a picture of a person's chromosome – taken and stained on a microscope slide.

Why do chromosomes look like this?

Chromosomes are *very small* but can be *specially prepared* so we can *see them* using a mi*croscope*. *Chromosomes* are *best seen* during mitosis (cell division), when they are c*ondensed* into *fuzzy shapes* you *see* here. *Chromosomes* taken from *dividing cells* are a*ttached* to a *slide* and *stained* with a *dye* called Giemsa (pronounced JEEM-suh). This d*ye* gives *chromosomes* a *striped* appearance because it *stains* the *regions* of DNA that are rich in:

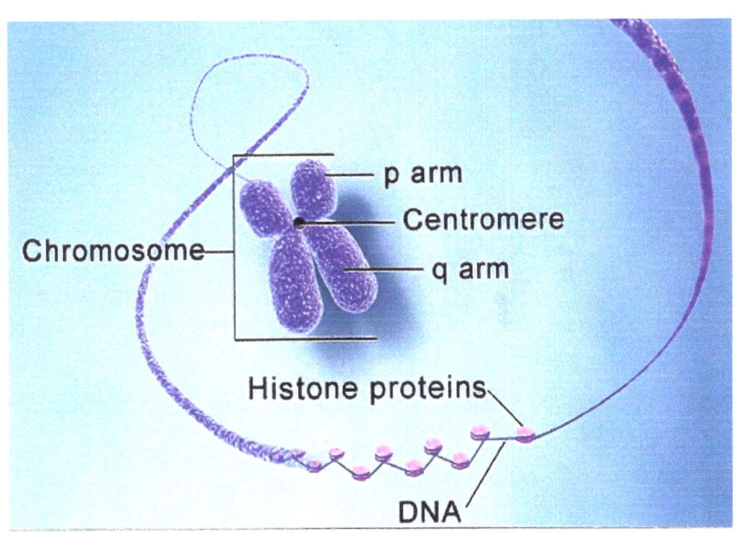

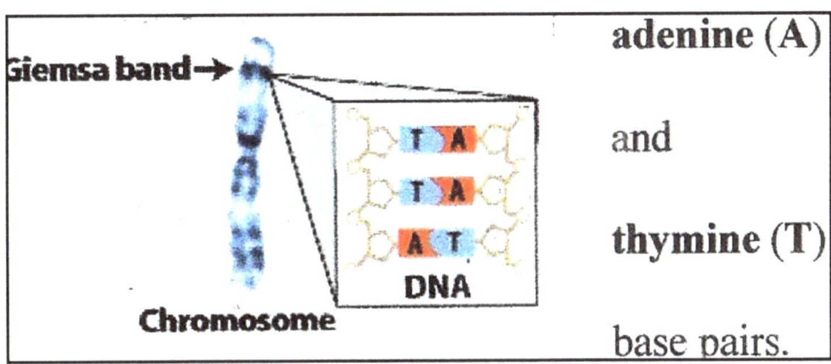

adenine (A) and thymine (T) base pairs.

INTRODUCING THE EVOLUTION FROM FISH TO MANKIND

Fish to Amphibians

Cambrian: 570-505 million years ago Vertebrates belong to the phylum **Chordata** – characterized by a cartilaginous notochord, a *rod-like* structure that serves to provide support for the animal body and protection for the dorsal nerve chord. The first Chordates are noted in the Cambrian period (i.e., the **Burgess Shale**). One lineage of Chordates will *evolve* into the sub-Phylum **Vertebrata** – the **vertebrates**.

Ordovician: 505-430 million years ago

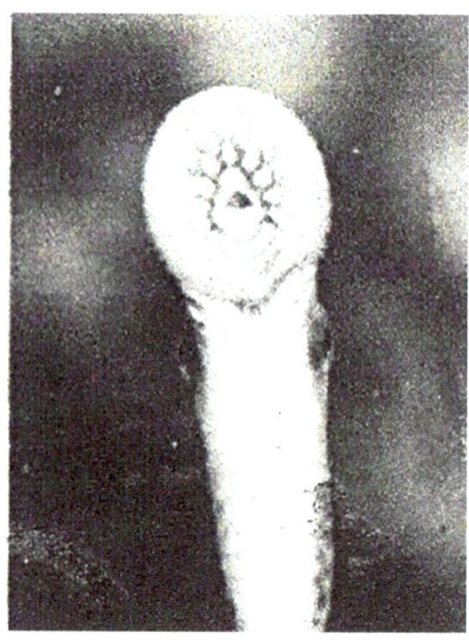

The first vertebrates were jaw-less fishes (Class Agnatha)

Modern *agnathans* include the *hagfish* and the *sea lamprey* (Petronmyzon), an *ectoparasite* of other fishes. A lamprey is shown in full body view above. The head is toward the right and the gill slits are evident. Instead of jaws, the lamprey has a *sucker-like* mouth. This mouth attached the lamprey to the side of its host. The *rasp-like* structures in the mouth "file" through the body wall of the host, permitting the lamprey to *feed* on the *body fluids* of the host.

Lobe-finned Fish and the Transition to Land

Crossoptergian or *lobe-finned s* (as opposed to regular *ray-finnedfishes*) are the *group* from which the **amphibians** evolved.

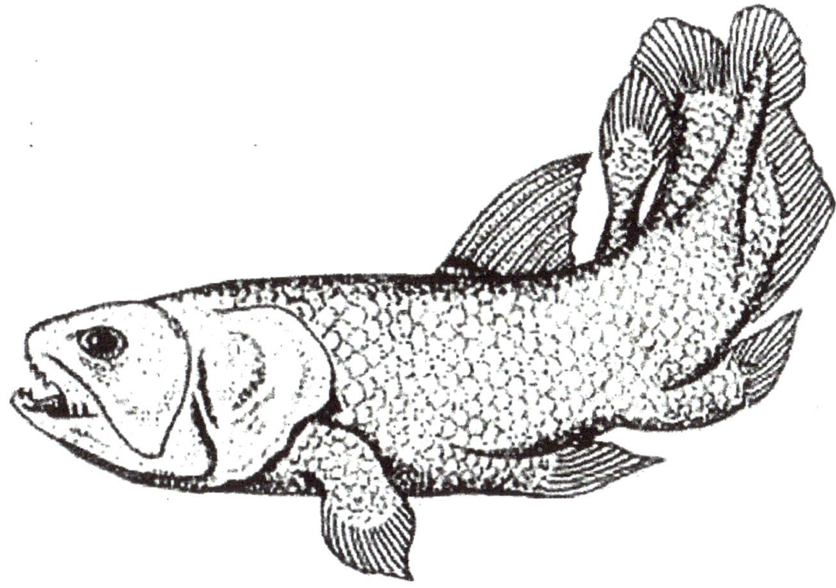

The lobe fin *structure*represents a pre-adaptation that facilitated the evolution of *tetrapodlimbs*. This is a *sketch* of theonly known living *lobe-finned fish*,

Latimeria

Latimeria, caught off the coast of South Africa in 1939.

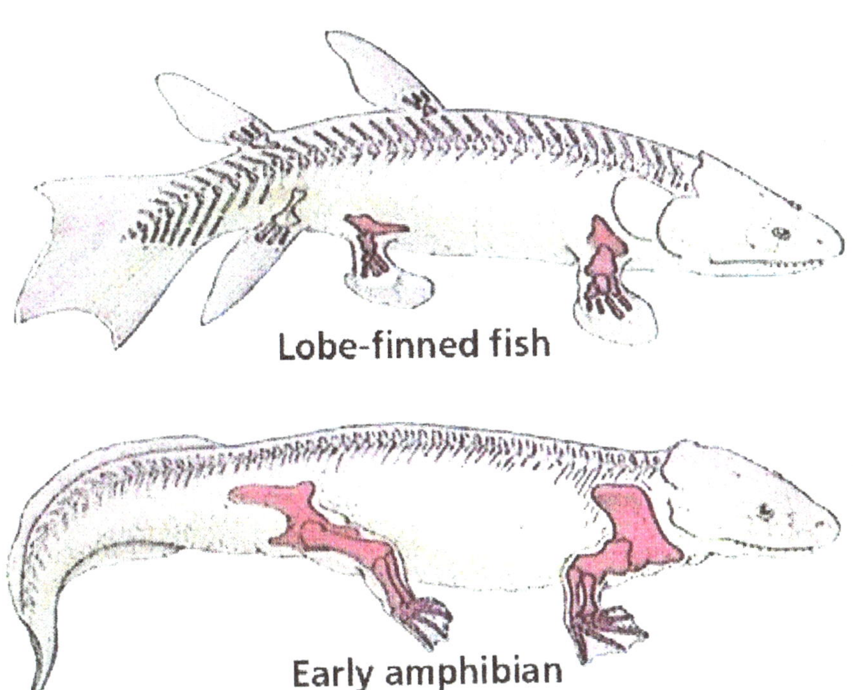

Comparisons of the skeletons of a crossotrygian lobe-finned fish and early amphibian image

Class Dunkleoteus

Devonian: 408-360 million years ago The Devonian, often referred to as the "age of fishes," was a time of the *explosive* diversification of fishes. The major *evolutionary innovation* was the development of *jaws*, derived from the *fusion* of *gill arches*. A group of fishes known as *Placoderms* (Class Placodermi) were the *first* to posses *jaws*, and are probably the *lineage* from which all other fishes evolved. *Placoderms* had a *cartilaginous* skeleton. Dunkleosteus (shown above) was a *large* (10 meter) *armored* Placoderm that exemplified the predatory life-style made possible by *jaws*. Most of the *armor* was concentrated in the *head region*, and the *fins* suggest that *this* fish could swim *well*. *Placoderms* went *extinct*, but *early* in the *Devonian* they gave *rise* to *two* classes of fish diversity: Class Chondrichthes and Class Osteichhtyes.

Class Chondrichthyes

Class Chondrichthyes

This class of cartilaginous fishes includes *Sharks, rays*, and *chimeras*. Since their origin, sharks have been the *predominant predators* in the *marine* environment. While sharks, as a group, trace their origin back to the *Silurian* (**438-408 million years ago**), they have been subject to multiple episodes of *adaptive radiation* – the last occurring in the *Mesozoic Era* (**245-66 million years ago**). In that sense, *sharks* and their *relatives* are as *modern* as any other fish in the *oceans* or *freshwaters*.

Class Osterichthyes

All the fish discussed to this time have *skeletons* of *cartilage*. The *Osteichthyes* evolved *bony* skeletons that assisted In their adaptive radiation into almost every *marine* and *freshwater* habitats.

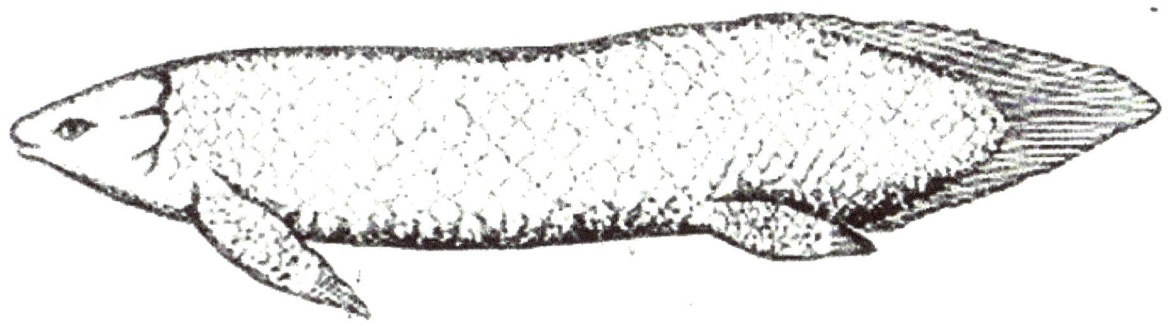

The **Teleost** *or* **ray-finned** *fishes are the most diverse of the two sub— groups and have an excellent fossil record in both marine and freshwater habitats.*

Vertebrate Chordates

The *vertebrates* comprise a *large group* of *chordates*, subdivided into *seven classes*: 3 fish classes, amphibians, reptiles, birds, and mammals.

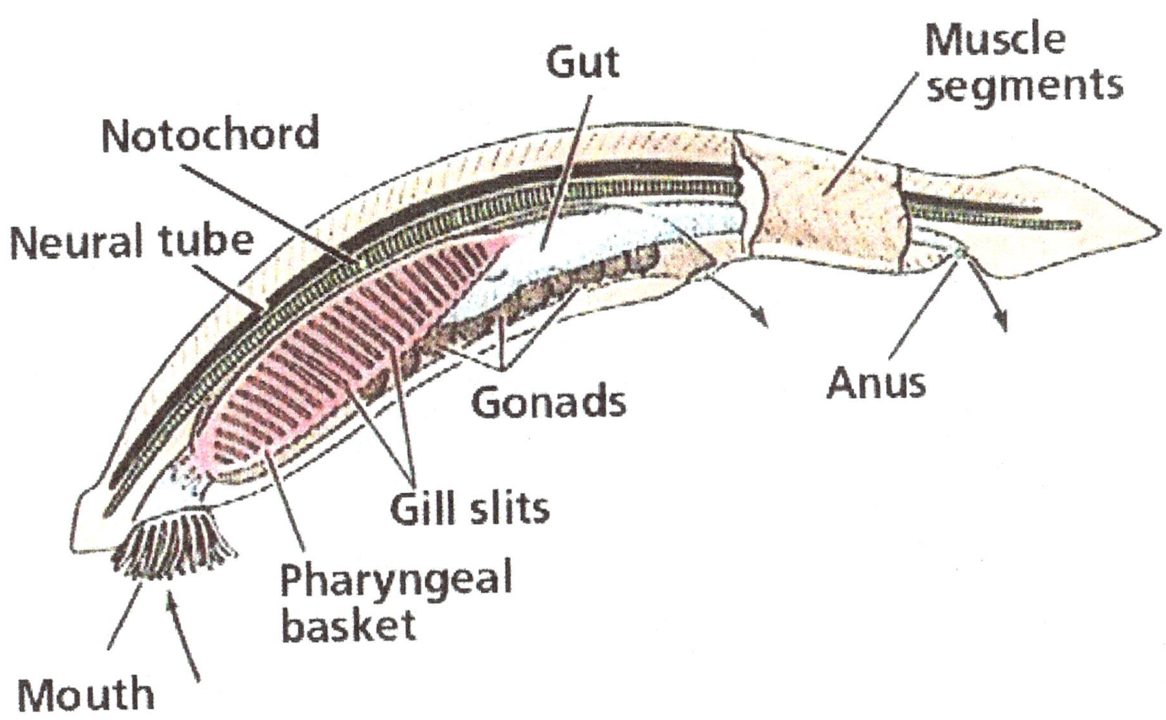

Anatomy of Ascaris, a typical chordate

This graph of the fossil record makes it easy to see the transformation of fish into amphibians

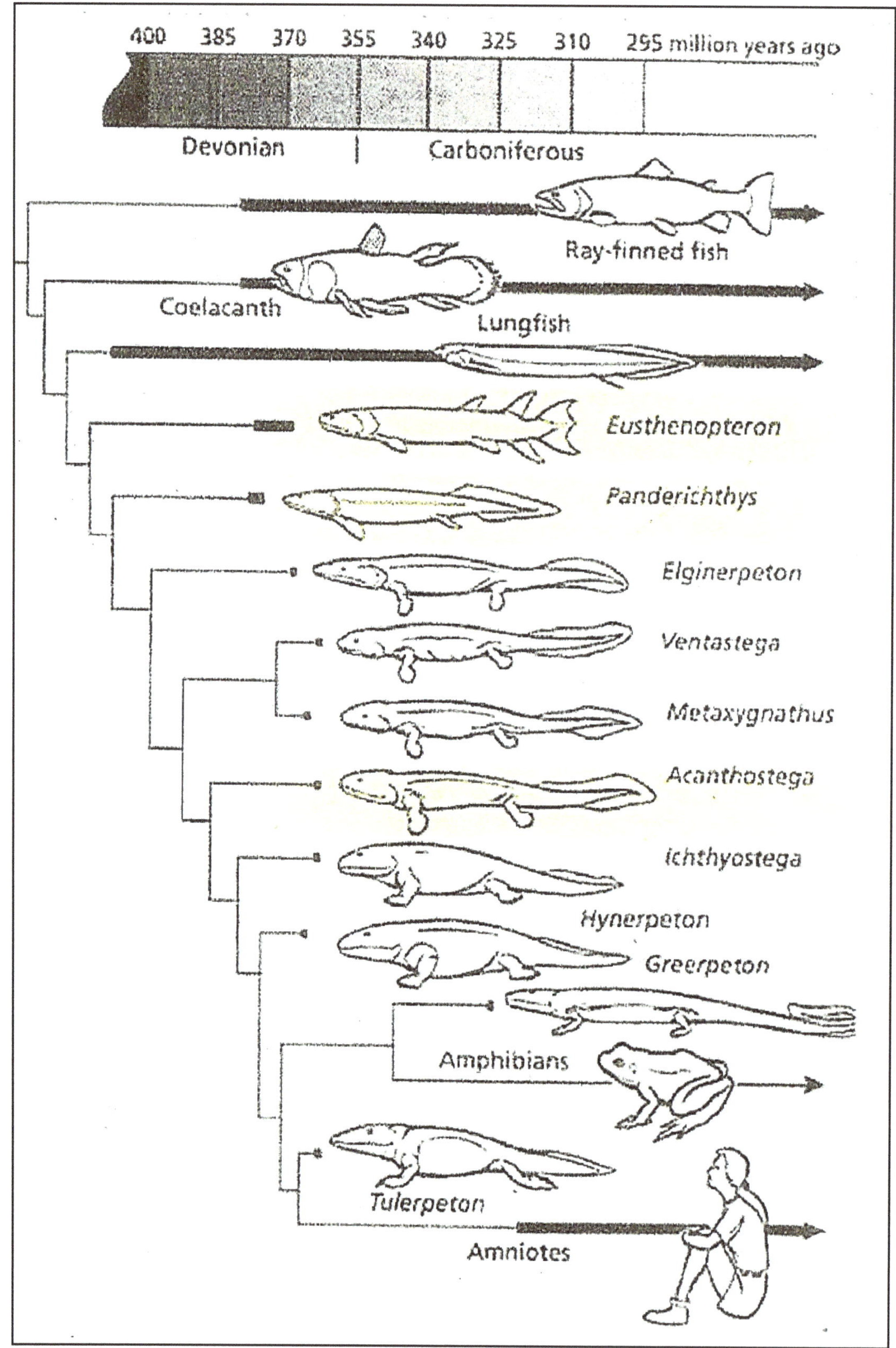

More About Lobe-Fins: Sarcoptergii

Fleshy Fins and Us Too
Lobe-fin fishes form one of the two known lineages of *bony fishes* (the **Osteichthyes**). *Ray-fin fishes* (**Actinopterygii**), which form the *other* lineage, are *arguably* the *most successful* of *vertebrates* and certainly the *most successful* "fishes." On the other hand, *lobe-fins* (**Sarcopterygii**) are *currently* represented only by the *coelacanth* (**Latimeria chaluminae**), and six species of lungfishes: *Lepidosiren paradoxa, Neoceratodus forsteri*, and four species of *Protopterus*. Lobe-fins experienced considerable success, nevertheless, during the *Paleozoic Era* (**570-245 million years ago**). They exhibited a greater *diversity* than the *ray-fins* during the *Devonian* (**408-360 million years ago**) and *Carboniferous* (**360-286 million years ago**), and they were typically the **top** *predators* in many of the *marine* and *freshwater habitats* they occupied. Moreover, one group of lobe- fins gave rise to the **tetrapods**, which have become the **other** most successful group of *vertebrates*. Strictly speaking, since *tetrapods* evolved from *lobe-fins*, **all** tetrapods – including **us** – are also **lobe-fins!**

Lobe-fins are characterized by their *fleshy* pelvic and *pectoral fins* with well developed *bones* and *muscles*. These fins join (or *articulate*) to the body *via* a single *bone* (*humerus* to the *shoulder* or pectoral *girdle*, and *femur* to the *pelvis*. In contrast, *ray-fin* fishes sport fins that contain several *rod-like* bones that articulate directly with the *pectoral* and *pelvic girdles*. Lobe-fins also have a *hinged braincase* together with a corresponding intra-cranial *joint* in the *skull roof*. This allows for a cranial kinesis (flexibility within the head) that provides additional *bite strength, a* feature that is *secondarily* lost in most *lungfish, tetrapods* and the *tetrapod's* closest *lobe-fin relatives*.

Fossil lobe ***fins*** first appear in the Early in the *Early Devonian* (408-360 million years ago), and *diversified* into several *major* groups by the Middle Devonian Lobe-fin *diversity* remained high during the *Late Devonian, Carboniferous* (360-286 million years ago) and *Early Permian* (286-245 million years ago), but declined significantly thereafter.

Diplurus (upper left; Triassic – 145-208 million years ago);
Easthenopter (upper right: Middle Devonian, and Holoptychius

The early tetrapods **were the f**irst vertebrates (amphibians) to *truly walk* the *land*. Before ***tetrapods*** existed ***vertebrates*** were confined to living in ***aquatic*** habitats. The *tetrapods* began their *conquest* of *land* in the *Paleozoic*, around 360 million years ago. *Tetrapods* are a *name* that we have given to anything with *four feet* (tetra=four, pod=feet). Therefore all *land dwelling* vertebrates can be considered *tetrapods*. When discussing the EARLY *tetrapods*, we are *referring* to very *primitive* (fish-like) *groups* that have *none* of the *specializations* of their *living* descendents: *amphibians, mammals, reptiles*—including *birds*. *Tetrapods,* whose closest *living* relatives are *lungfish*, include two *main* groups: **amphibians** and **amniotes**. The *amniotes* in turn have two main groups – the **synapsids** (including *mammals*) and the **sauropsids** (including reptile and their fossil relatives.

The *tetrapod* at left is an **anthracosaur** called **Seymouria**, from the UCMP collections. It is from North America, and is thought to be from the Permian period, between **286 and 248 million years ago**. It was previously thought to be related to *amniotes*, but recent *studies* suggest that it is a **stem-tetrapod**.

Tetrapod

Tetrapod (Greek *tetrapodis*, Latin *quadruped*, "four-legged") are vertebrate animals having four feet, legs or leg-like appendages. Amphibians, reptiles, dinosaurs, and mammals are all tetrapods, and even the *limbless* snakes are tetrapods by *descent*. The earliest tetrapods radiated from the **Sarcoptergii**, or *lobe-finned fish*, into *air-breathing amphibians* in the *Devonian* period.

Devonian tetrapods

Research by *Jennifer A. Clack* and her colleagues showed that the *earliest* tetrapods, such as *Acanthoste* tetrapods were *wholly* aquatic and quite *unsuited* to life on land. This overturned the earlier *view* that fish *invaded* the land – either in search of prey (like modern *mudskippers*) or to find water when the pond they lived in dried out – and later evolved legs, lungs, etc.

The *first* tetrapods are now thought to have evolved in *shallow* and *swampy freshwater* habitats, towards the end of the *Devonian*, a little more than **365 million years ago**. By the late Devonian, land plants had *stabilized* freshwater habitats, allowing the first wetland ecosystems to develop, with increasingly complex food webs that afforded new opportunities, (1) Freshwater habitats were not the only places to find water filled with organic matter and *choked* with plants with dense vegetation near the water's edge.

Swampy habitats like shallow wetlands, coastal lagoons and large brackish river deltas also existed at this time, and there is much to suggest that this is the *kind* of environment in which the tetrapods evolved. Early fossil tetrapods have been found in marine sediments, and because fossils of primitive tetrapods in general are found scattered all around the world, they have *spread* by following the *coastal lines* – they could not have lived in freshwater only.

The common ancestor of all present **gnahostomes** lived in freshwater, and later migrated back to sea. To deal with the much *higher* salinity in *sea* water, they evolved the ability to *turn* the *nitrogen* waste product *ammonia* into harmless *urea*, storing it in the body to make the *blood* as *salty* as the sea water without *poisoning* the organism. Ray- finned fishes later *returned* to freshwater and *lost* this ability. Since their *blood* contained more *salt* than *freshwater*, they could simply get *rid* of ammonia through their *gills*.

When they finally *returned* to the sea again, they could *not* recover their old trick of *turning* ammonia to *urea*, and they had to evolve *salt excreting glands* instead.

Tetrapods: Fossil Record

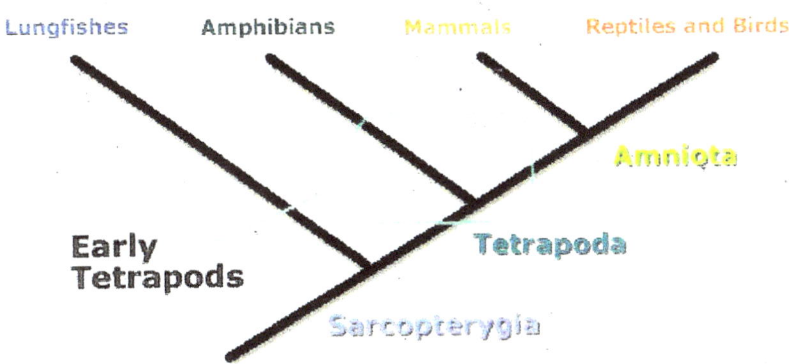

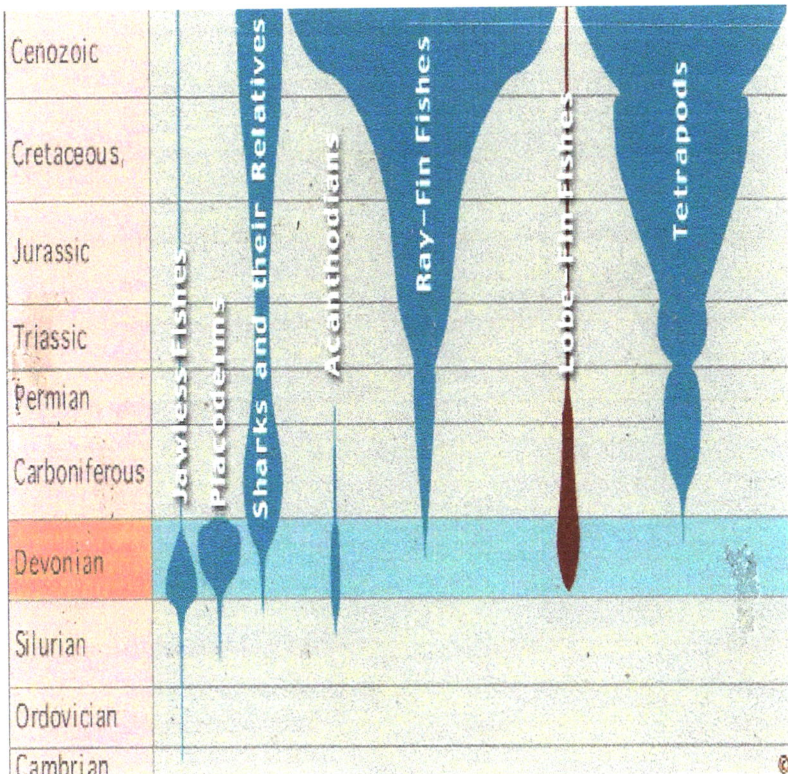

Lungfishes do the *same* when they are *living* in the water, making *ammonia* and no *urea*, but when the water *dries up* and they are forced to *burrow* down in the mud, they *switch* to *urea* production. Like *cartilaginous* fishes, the *coelacanth* were able to *store* urea in their *blood*, and the only known amphibians that lived for *long* periods of time in salt water (presently, the toad *Bufo marinus* and the frog *Rana cancrivora*). These are traits they have *inherited* from their *ancestors*.

If early tetrapods *lived* in freshwater, and *lost* the ability to *produce* urea and used ammonia *only*, they would have had to evolve it from *scratch* again later. Not a *single* species of all the ray-finned fishes *living* today had been able to do that, so it is *not* likely the tetrapods would have done so either. Terrestrial animals that can *produce* ammonia o*nly* would have to *drink* constantly, making a life on land *impossible* – a few exceptions exist, so some terrestrial *woodlice,* for example, can *excrete* their nitrogenous waste as ammonia *gas*. This probably was also a problem at the start when the tetrapods started to spend time *out* of water, but eventually the urea system would *dominate* completely.

Because of this it is *not likely* that they still *hadn't* forgotten how to make urea. Even if some *never* went to land (or extinct primitive species that returned to water), they nevertheless would have *adapted* to freshwater lakes and rivers *again*.

Primitive tetrapods *developed* from a *lobe-finned fish* (an "**osteolepid sarcopterygian**"), with a two-lobed *brain* in a flattened *skull*, a wide *mouth* and a short *snout*, whose upward-facing *eyes* show that it was a bottom-dweller, and had already developed adaptations of *fins* with *fleshy* bones – the "living fossil" **coetacanth** is a related marine lobe-finned fish *without* these shallow-water adaptations.

Panderichthys

Even more closely related was Panderichthys, which even had a choana. These fishes used their fins as *paddles* in shallow-water habitats choked with plants and detritus.

Their fins could have also been used to *attach* themselves to *plants* or similar objects while they were laying in wait to ambush for prey. The universal tetrapod characteristics of front limbs that bend *backward* at the elbow and hind limbs that bend *forward* at the knee can plausibly be traced to early tetrapods living in shallow water.

It is now clear that the common ancestor of the bony fishes had a primitive *air- breathing* lung (later *evolved* into a *swim* bladder in most ray-finned fishes). This suggests that it evolved in *warm* waters, the kind of habitat the lobe finned fishes were living and made use of their simple lung when the oxygen level in the water became too low.

The lungfishes are now considered as being the *closest* living *relatives* of the tetrapods, even closer than the *coelacanth*.

Fleshy lobe fins supported tetrapods on *bones* rather than ray-stiffened *fins* seems to have been an original trait of the bony fishes (*Osteichthyes*). The lobe-finned ancestors of the tetrapods evolved them further, while the ancestors of the ray-finned fishes (*Actinopteryii*) fishes evolved their fins in the opposite direction. The most primitive group of the ray-fins, the *bichirs*, still have fleshy frontal fins.

The *first* Devonian tetrapod identified from Asia was recognized from a fossil jawbone reported in **2002**. The Chinese tetrapod *Sinostego pani* was discovered among fossilized tropical plants and lobed-finned fish in the red sandstone sediments of the *Ningxia Hui*, autonomous region of northwest China. This finding subsequently extended the geographical range of these animals and has raised new questions about the worldwide distribution and great taxonomic diversity they achieved within a relatively short time.

These *earliest* tetrapods were *not* terrestrial. The *earliest* confirmed terrestrial forms are known from early *Carboniferous* deposits, some **20 million years** later. Still, they may have spent very *brief* periods out of water and would have used their legs to *paw* their way through the mud.

Why they went to land is the first place still *debated*. One reason could be that the *juveniles* who had completed their *metamorphosis* had what it took to make *use* of what land had to *offer*. They already adapted to *breath* air and move around in *shallow* waters *near* land as a protection – just as modern fish and amphibians often spend the first part of their lives in the comparative safety of shallow waters like mangrove forests.

Two very *different* niches partially *overlapped* one another, with the *juveniles* in the diffuse line between. *One* of them was overcrowded and dangerous while the *other* was much *less* crowded and safer, offering *less* competition over resources. The terrestrial niche was a much *more* challenging place for primarily *aquatic* animals, but because of the way evolution and the selection pressure work, those juveniles who could take advantage would be rewarded, in that once they gained a *small* foothold on land, *evolution* would take care of rest, thanks to all of their pre-adaptations along with being at the *right* place at the *right* time.

At this time there were a lot of invertebrates crawling around on land and nearby water, in moist soil and wet litter, more than big enough to give the small ones a good meal. Some were even big enough to *eat* small tetrapods, but nevertheless, land would still be a much better place and offer more than the waters **if** they knew how to make use of it.

Adults would be too *heavy* and *slow* and demand bigger prey. Juveniles were much lighter, faster and were satisfied with relatively small invertebrates. Mudskippers of today are able to *snap* insects in flight while on land – if so, then the same can similarly be said of the early juvenile tetrapods.

Initially making only tentative forays onto land, as the generations went by, however, they adapted to terrestrial environments, spending *longer* periods *away* from the water, along with a spending a *longer* part of their childhood on land before *returning* again to the water for the remainder of their lives. It is also possible that the adults started to spend some time on land as the skeletal modifications in early tetrapods such as **Ichthyostega s**uggest, but only to bask in the sun *close* to the water's edge, though not to *hunt* or move. It is a fact that the first *true* tetrapods adapted to terrestrial locomotion were *small*, but progressively *increasing* in size with time.

Ichthyostega

As is the case with all living things, grown tetrapods *retained* most of their anatomical and other forms of adaptations *gained* from their juvenile stage, giving them *modified* limbs along with other traits of needed terrestrial properties. To be *successful* adults they had to first be *successful* juveniles. The adults of some of the *smaller* species probably, were in *that* case, also able to move on land when sufficiently evolved.

If some sort of *neoteny* or *dwarfism* occurred, making the animals sexually mature and fully grown while still living on land, they would need *only* to visit and *drink* the water to reproduce.

Carboniferous tetrapods (360-286 million years ago)
Until the 1990s, there was a *30 million year* gap in the fossil record *between* the late Devonian tetrapods and the *reappearance* of tetrapod *fossils* in recognizable mid-*Carboniferous* amphibian lineages. It was referred to as "**Romer's Gap**", named after the *Palaeontologist* who recognized it.

During the "gap," tetrapod *backbones* developed, as did *limbs* with digits and other adaptations for terrestrial life. Ears, skulls and vertebral columns all underwent *changes* too. The *number* of digits on hands and feet became standardized at **five**, as lineages with more digits *died* out. The very *few* tetrapod fossils in the "gap" make them all the *more significant* and *precious*.

The *transition* from an aquatic lobe-finned fish to an *air-breathing* amphibian was a MOMENTUS occasion in the evolutionary history of the vertebrates. For an animal to *live* in a **gravity-neutral**, aqueous environment and then *invade* one that is e**ntirely different** required *major* changes to the overall body plan, both in *form* and in f*unction*. **Eryops** is an example of an animal that *made* such adaptations. It *retained* and *refined* most of the traits found in its fish ancestors. Sturdy limbs *supported* and t*ransported* its body while out of water. A thicker, stronger backbone *prevented* its body from **caving** under its own weight. By also *utilizing* vestigial fish jawbones, a r*udimentary* ear was developed, allowing *Eryops* to hear *airborne* sound.

Eryops

By the *Visean* age of *mid-Carboniferous* times the early tetrapods had *radiated* into at least *three* main branches: (1) *temnospondyls*, (2) *anthracosaurs*, and (3) b*abhetis*. Recognizable basal-group tetrapods are representative of the t*emnospondyls* (e.g. *Eryops*) and similarly primitive *anthracosaurs*, which were the relatives and ancestors of the *Amniota*. *Depending* on whatever *authority* one follows, modern amphibians (*frogs, salamanders* and *cacilians*) are *derived* from *one* or the *other* (or probably *both*, although this is now a *minority* position) of these *two* groups. The first a*mniotes* are known from the *early* part of the *Late Carboniferous*, and during the *Triassic* counted among their number the *earliest* mammals, turtles, and crocodiles (lizards and birds appeared in the *Jurassic*, and snakes in the *Cretaceous*). As living members of the tetrapod clan – that is, of the tetrapod "crown-group" – these varied tetrapods represent the *phylogenetic end-points* of these two *divergent* lineages. The t*hird*, more primitive, Carboniferous group, the *baphetids*, left *no* modern survivors. The *Lepospondyli* are an extinct Palaeozoic group, of uncertain relationships.

In the *Permian Period*, as *separate* tetrapod lineages each *developed* their own unique way, the term "*tetrapods*" became *useful*. In addition to *temnopondyl* and *anthracosur clades*, the *early* "amphibia" (**labyrinthodons**), there were *two* important divergent clades of *aminiotes*, the **Sauropsida** and the **Synapsids**, of which the latter were the most important and successful *Permian* animals. Each of these lineages, however, *remains* grouped with the *tetrapods*, just as *Homo sapiens* could be considered a very highly- specialized kind of *lobe-finned fish*.

Living tetrapods

There are three main categories of living ("crown group") tetrapods, all of which also include many extinct groups:

Amphibia: frogs and toads, newts, salamanders, and caecilians *Sauropsida*: reptiles, dinosaurs, and birds *Synapside*: mammals Note that *snakes* and other *reptiles* are considered tetrapods because they are *descended f*rom ancestors who had a *full* complement of **limbs**. Similar considerations apply to c*aecilians* and *aquatic* mammals.

Classification

All early tetrapods and *tetrapodomorphs* that were *not* true *amphibians* or *amniots* were once placed together in the *paraphyletic* group *Labyrinthodhonts,* were distinguished mainly by their complex *dentine or* [infolding] *teeth* structure, a feature *shared* with c*rossopterygian* fish. The **labyrinthodonts** were *divided* into the *Ichthyosgalia* (another p*araphyletic)* assemblage of *primitive* tetrapods and kin, such as *Ichthhyostega*, the *Temnospondyli* (possibly members of *Amphibia*), and the *Anthbracosauria, close* relatives of amniotes. The *main* difference between the *three* groups is based on their respective *vertebral* structures. The Anthracosaurias had *small pleurocentra*, which *grew* and *fused*, becoming the *true* centrum in *later* vertebrates. In *contrast*, the Temnospondli had a *conservative* vertebral column in which the pleurocentra remained *small* in p*rimitive* forms, *vanishing* entirely in the more *advanced* ones. The *intercentra* are *large* and form a *complete* ring.

Although the temnospondyls flourshed in many forms in the Late *Paleozoic* and *Traissic*, they were an *entirely* self-contained group and did *not* give rise to any later tetrapod groups. It was the *sister* group **Anthracosautia** that *gave* rise to *reptiles*.

Apartial taxonomy of the tetrapods:

Phylum Chordata:

* Class Sarcopterygli
    * **Subclass Tetrapodomorpha**
    * Eusthenopteron
        * *Panderichthys*
            * *Tiktaalik*
                * **Subclass tetrapods**
                    * Family Elginerpetontidae
                    * Family Acanthostegidae
                    * Family Ichthyostegidae
                    * *Hynerpeton*
                    * Family Whatcheeriidae

* Family Crassigyrinidae
* Family Loxominatidae
* Batrachomorpha
    * **Class Amphibia – Amphibians**
        * Subclass Lepospondyli
            * Subclass Temnospondyli
            * Subclass Lisamphibia – frogs

              Salamanders

              Reptiliomorpha
                * **Amniota**
                    * Class Sauropsida – Reptiles
                    * Class Aves – Birds
                * Class Synapsida – Mammal-like
                    * Class Mammalia –

                      Mammals

Ampelosaurus

Anatomical features of early tetrapods

The tetrapod's ancestral fish must have passed *similar* traits to show inheritance by early tetrapods, including *internal* nostrils (to *separate* the **breathing** and **feeding** passages) and a large *fleshy* fin built on bones that could give rise to the tetrapod *limb*. The r*hipidiatian* **crossopterygians** fulfill every requirement for the ancestry. Their *palatal* and *jaw* structures were *identical* to those of *early* tetrapods, and their *dentition* was identical to those of *early* tetrapods, their dentition was *identical* too, with *labyrinthine* teeth *fitting* in a pit-and-tooth arrangement on the palate. The *crossopterygian paired* fins were *smaller* than tetrapod limbs, but the skeletal structure was very *similar* in that the crossopterygian had a *single* proximal bone (analogous to the *humerus* or *femur*), *two* bones in the next segment (forearm or lower leg), and an *irregular* subdivision of the fin, roughly *comparable* to the structure of the *carpus/tarus* and phalanges of a **hand**.

The *major* differences between crossoptergians and *early* tetrapods was in relative development of *front* and *back* skull portions; the *snout* is much *less* developed than in most *early* tetrapods and the post-orbital skull is exceptionally *longer* than an amphibian's.

A great many kinds of *early* tetrapods lived during the *Carboniferous* period. Their ancestor, therefore, would have lived *earlier*, during the *Devonian Period.* Devonian **Ichthyostegids** were the *earliest* of true tetrapods, with a skeleton that is *directly* comparable to that of *rhipidstian* ancestors. Early *termnospondyls* (Late Devonian to Early Mississippian) still had some *ichthyostegid* features such as *similar* skull bone patterns, *labyrinthine* tooth structure, the *fish* skull-hinge, pieces of *gill* structure *between* the *check* and *shoulder*, and the *vertebra* column. They had *lost* several other fish features, however, such as the *fin rays* in the tail.

In order to *propagate* in the *terrestrial* environment, *certain* challenges had to be o*vercome*. The animal's body needed *additional* support, because *buoyancy* was no longer a *factor*. A *new* method of *respiration* was required in order to *extract* atmospheric oxygen, instead of oxygen *dissolved* in *water*. A means of *locomotion* would need to be developed to *traverse* distances *between* waterholes. Water *retention* was now *important* since it was *no* longer the *living* matrix, and it could be *lost* easily to the environment.

New sensory *input* systems were finally *required* if the animal was to have *any* ability to f*unction* reasonably while *on land.*

Skull

The most *notable* characteristic that *make* a tetrapod skull *different* from a fish's *are* the relative *frontal* and *rear* portion lengths. The fish had a *long* rear portion while the *front* was *short*; The *orbital* vacuities were thus located towards the *anterior* end. In the tetrapod, the *front* of the skull *lengthened*, positioning the orbits *back* on the skull. The lacrimal bone was *not* in contact with the *frontal* anymore, having been *separated* from it by the *prefrontal* bone. Also of *importance* is that the skull was now free to *rotate* from s*ide* to *side*, independent of the spine, on the *newly* forming **neck**.

A diagnostic character of *temnospondyls* was that the *tubular* bones (which formed the posterior corners of the skull-table) were *separated* from the respective *left* and *right* parietals by a *sutural* junction *between* the *postparietals* and *supratemporals.* Also at the r*ear* of the skull, all bones dorsal to the *clethrum* were *lost.*

The *lower* jaw of, for example, *Eryops* resembled its *crossoptergian* ancestors in that on the *outer* surface lay a *long* dentary that *bore* teeth. There were also bones *below* the dentition *on* the jaw: two splenials, the angulary and the surangular. On the *inside* were usually *three* coronoids that bore teeth and lay *close* to the dentition. On the *upper* jaw was a row of *marginal* labyrinthine teeth, located on the maxilla and premaxilla. In *Eryops*, as all early amphibians, the teeth were *replaced* in *waves* that traveled from the f*ront* of the jaw to the *back* in such a way that *every* tooth was mature, and the *ones* in between were young.

Dentition

The "labyrinthho[donts]" had *peculiar* tooth structures from which their names were d*erived* and, although not exclusive to the group the labyrinthine dentition is a *useful* indicator as to *proper* classification. The *important* feature of the tooth is that the enamel and dentine were *folded* in such a way as to form a complicated *corrugated* pattern when viewed in cross-section. This *in-folding* resulted in the *strengthening* of the teeth and thus increased wear-resistance. Such teeth survived for **100 Million years**, first among c*rossoptery*, then stem reptiles.

Modern amphibians *no longer* have this type of dentition but rather *pleurodont* teeth, in *fewer* numbers of the whole group.

Sensory organs

There is a density *difference* between *air* and *water* that causes *smells* (certain chemical compounds detectible by *chemoreceptors*) to *behave* differently. An animal first venturing out onto land would have difficulty in *locating* such chemical signals if its *sensor* apparatus was designed for *aquatic* detection.

Fish have a *lateral* line system that *detects* pressure fluctuations *in* the water. Such pressure is *non-detectable* in air, but *grooves* for the lateral line sense organs were found on the *skull* of *labyrinthodonts*, suggesting a *partially* aquatic habitat. *Modern* amphibians, which are *semi-aquatic*, exhibit this feature whereas it has been retired by h*igher* vertebrates. The *offactory opithhelium* would also have to be *modified* in order to detect *airborne* odors.

In addition to the *lateral line* organ system, the *eye* had to *change* as well. This c*hange* came about because the *refractive* index of light *differs* between *air* and *water*, so the *focal* length of the lens was *altered* in order to properly function. The eye was now e*xposed* to a relatively *dry* environment rather than being *bathed* by water, so *eyelids* developed and *tear ducts* evolved to produce a liquid, *moistening* the eyelid.

Hearing

The *balancing* function of the *middle-ear* was *retained* from *fish* ancestry, but delicate air vibrations could *not* set up *pulsations* through the *skull* in order for it to function a proper *auditory* organ. Typical of most lab *labyrinthodonts*, the spiracular gill pouch was *retained* as the *otic* notch, *closed in* by the *tympanum,* a thin, tight membrane.

The *hyomandibula* of fish migrated *upwards* from its jaw-supporting position, and was *reduced* to form the *stapes*. Situated between the *tympanum* and *braincase* in an air-filled cavity, the staple was now capable of *transmitting* vibrations from the exterior of the head to the interior. Thus the *stapes* became an *important* element in an impedance matching system, *coupling* airborne sound waves to the *receptor* system of the *inner* ear. This system had evolved *independently* within several different amphibian lineages.

"Did Our Ancestors *Breathe* through Their Ears?"

Girdles

The *pectoral* girdle of early tetrapods such as *Eryops* was *highly* developed, with a *larger* size for both *increased* muscle attachment to it and its *limbs*. Most notably, the *shoulder* girdle was *disconnected* from the *skull*, resulting in *improved* terrestrial locomotion. The **crossopterygian** *clethrum* was *retained* as the clavicles, and the *interclavicle* was well- developed, lying on the *underside* of the chest. In primitive forms, the two clavicle, and the interclavicle could have grown ventrally in such a way as to *form* a broad chest plate, although such was *not* the case in *Eryops*. The *upper* portion of the girdle had a *flat*, scapular blade, with the glenoid cavity situated *below* performing as the articulation surface for the humerus, while ventrally there was a large, flat coracoid plate turning in toward the *midline*.

The pelvic girdle was also much *bigger* than the simple plate found in fishes, accommodating more muscles, it extended *far* dorsally and was *joined* to the backbone by *one* or *more* specialized *sacral* ribs. The *hind* legs were somewhat specialized in that they not only *supported* weight, but also *provided* propulsion. The dorsal extension of the pelvis was the *ilium,* while the broad ventral plate was comprised of the pubis in *front* and the ischium in *behind*. The *three* bones met at a *single* point in the *center* of the pelvic triangle called the *acetobulum,* providing a surface of *articulation* for the femur.

The *main* strength of the ilio-sacral attachment of *Eryops* was by *ligaments*, a condition structurally, but not phylogentically, intermediate between *that* of the *most* primitive *embollomerous amphuibians* and *early* reptiles.

The condition that is more usually *found* in *higher* vertebrates is that *cartilage* and *fusion* of the sacral ribs to the blade of the ilium are *utilized* to ligamentous attachments.

Limbs

The humerus was the *largest* bone of the arm, its head articulating with the glenoid cavity of the pectoral girdle, distally with the radius and ulna. The radius resided on the *inner* side of the forearm and rested directly *under* the humerus, *supporting* much of the weight, while the *ulna* was located to the *outside* of the humerus. The ulna had a head, whose muscles pulled on to extend the limb, called the *olecranon* that extended above the edge of the humerus.

The radius, the ulna were *articulated* with the carpus, which was a proximal row of these elements: the radiale underlying the radius, the ulnare *underneath* the ulna and the intermediate *between* the two. A large *central* element was *beneath* the *last* and may have *articulated* with the radius. There were also *three* smaller Centralia lying to the radial side. Opposite the head of each toe lay a series of five carpals. Each digit had a first segment, the metacarpal, lying in the region.

The pelvic limb bones were essentially the *same* as in the *pectoral* limb, but with different names. The *analogue* to the humerus was the *femur*, which was *longer* and *slimmer*. The two lower arm bones corresponded to the tibia and fibula of the hind leg the former being the innermost and the outermost bones. The tarsus in the hind version of the carpus and its bones *corresponded* as well.

Feeding

Early tetrapods had a *wide* gaping jaw with *weak* muscles to open and close it. In the jaw were *fang-like* palatal teeth that, when *coupled* with the gape, suggests an inertial feeding habit. This is when the amphibian would *grasp* the prey and, *lacking* any chewing mechanism, *toss* the head up and *backwards*, throwing the prey further back *into* its mouth, feeding such is seen today in the *crocodile* and *alligator*.

Cynognatnus

The *range* of modern adult *amphibians* is quite *fleshy* and attached to the *front* of the lower jaw, so it is reasonable to speculate that it was fastened in a *similar* fashion in p*rimitive* forms, although it was probably *not* specialized as it is in a *frog*.

It is assumed that early adult tetrapods were *not* very active, thus a predatory lifestyle was probably *not* the norm. It is more likely that it fed on *fish* either *in* the water or *on* those that became stranded *at* the margins of lakes and swamps. A large supply of terrestrial invertebrates were also abundant at the time, which no doubt have provided a fairly adequate food supply.

Respiration
Today's amphibian breathes by *inhaling* air into their lungs, where *oxygen* is absorbed.

They also breathe through the moist lining of the mouth and skin. *Eryops* also inhaled, but its ribs were *too* closely *spaced* to suggest that it did this by *expanding* the rib cage.

More likely, it *opened* its mouth and nostrils, depressing the hyoid apparatus to *expand* the oral cavity, closed its mouth and nostrils finally and elevated the floor of the mouth to force air back into the lungs – in other words, it *gulped* then *swallowed*. It probably e*xhaled* by contraction of the elastic tissue in the lung walls. Other special respiratory methods probably existed.

Circulation
Early tetrapods most likely had a *three-chambered* heart, as do *modern* amphibians and reptiles, in which oxygenated blood *from* the lungs and *de-oxygenated* blood from the respiratory tissues entered by separate atria, and was directed via a spiral valve to the appropriate vessel – *aorta* for oxygenated blood and *pulmonary vein* for deoxygenated blood. The spiral valve was *essential* to keeping the *mixing* of the two types of blood to a minimum, enabling the animal to have *higher* metabolic rates, to become more active than it would have otherwise had been.

Locomotion
Typical early tetrapod posture was by the *upper* arm and *upper* leg extending nearly *straight* out *from* the body. While the forearm and the lower leg *extended* d*ownward* from the upper segment at a night angle. The body weight was *not* centered o*ver* the limbs, but rather was transferred 90 degrees *outward* and *down* through the l*ower* limbs, which contacted the ground. Most of the animal's *strength* was used to just e*levate* its body *off* the ground for *walking*, which was probably *slow* and *difficult*. With this *sort* of posture, *only* short, broad strides could be achieved. This has been confirmed by fossilized footprints found in Carboniferous rocks.

Ligamentous attachments within the limbs were present in *Eryops*, being important because they were the *precursor* to *bony* and *cartilaginous* variations soon in terrestrial animals that use their *limbs* for *locomotion*.

Of all body parts, the **spine** was the most affected by the move from *water* to *land*. It had to *resist* the *bending* caused by body weight and had to *provide* mobility where needed. It was able to *bend* along its *entire* length previously. Likewise, the *paired* appendages had not *formally* been *related* to the spine, but the *slowly* strengthening limbs t*ransmitted* their support to the *axis* of the body.

INTRODUCING THE TRILOBITES?

Trilobites are remarkable, hard shelled, segmented creatures that existed over *300 million years ago* in the Earth's ancient seas. They went *extinct* before dinosaurs even came into the *picture*, and were *one* of the *key* signature creatures of the *Paleozoic Era*, the first to exhibit a proliferation of the *complex life-forms* that established the foundation of life as it is today. Although dinosaurs are the *most* well-known fossil life forms, trilobites are also a favorite among those familiar with *Paleontology* (the study of the development of life on Earth), and are found in the rocks of all continents.

Ancient Arthropods

Trilobites were among the early *arthropods*, a phylum of *hard-shelled* creatures with multiple *body segmentsand jointed legs* – although the legs, antennae and other finer structures of trilobites are *rarely* preserved. They *constitute* an *extinct class of athropods*, the *Trilobite*, made up of *ten orders*, over **150** *families*, about **5,000** *genera*, and over**17,000** *described species*. New species of trilobites are *unearthed* and *described* every year. This makes trilobites the *single* most *diverse group* of *extinct* organisms, and within the generalized *body plan* of trilobites there was a great deal of *diversity* of size and form. The *smallest* known trilobite species is just under **a millimeter long**, while the l*argest* include species from *over* **30** to **70** cm in length (roughly a foot to over two feet long!) With such a diversity of species and sizes, speculation on the *ecological* role of trilobites includes *planktonic, swimming*, and *crawling* forms, and can be presumed to have filled a varied set of *trophic (feeding) niches*, although perhaps mostly as d*etritivores, predators*, or *scavengers*. Most trilobites were about an **inch long**, and part of their appeal is that you could *hold* and *examine* an entire fossil animal and *turn* it a*round* in your *hand*. Not an *option*, needless to say, with your average *dinosaur*.

The Trilobite Body Plan

Whatever their size, all trilobite fossils have the *same* body plan, made up of *three* main body parts: a *Cephalon* (head), a segmented *thorax*, and a *pygidium* (tail piece).

The name "trilobite," which means "***three lobed***," however, is not in reference to those three body parts mentioned above, but to the fact that all trilobites bear a long central a***xial lobe***, flanked on each side by right and left ***pleural lobes*** (***pleu:*** side, rib). These t***hree lobes*** that run from the ***cephalon*** to the ***pygidium*** are what give ***trilobites*** their name, and are ***common*** to all trilobites ***despite*** their great diversity of size and form.

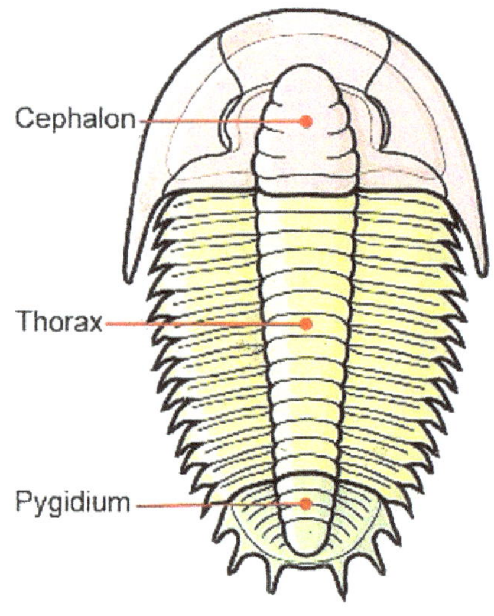

THE TRILOBITE BODY PLAN

Whatever their size, all trilobite fossils have a similar body plan, being made up of three main body parts: a **cephalon** (head), a segmented **thorax**, and a **pygidium** (tail piece) as shown at left. However, the name "trilobite," which means "**three lobed**," is not in reference to those three body parts mentioned above, but to the fact that all trilobites bear a long central **axial lobe**, flanked on each side by right and left **pleural lobes** (pleura = side, rib). These three lobes that run from the cephalon to the pygidium are what give trilobites their name, and are common to all trilobites despite their great diversity of size and form. You can examine the trilobite body plan in more detail using the links on the navigation bar below, or link directly to a page describing trilobite **major features**.

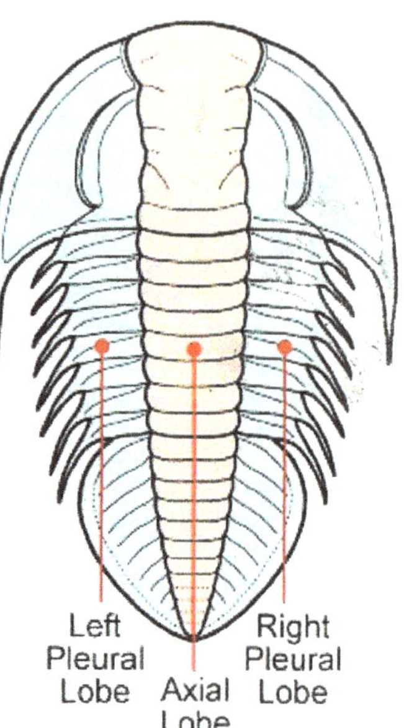

Trilobite Systematic Relationships

Primitive or advanced

For a long time, *trilobites* were considered the *most* primitive of *arthropods* since they were *among the first Cambrian arthropods discovered*, but this turns out to have been p*remature;* instead it was an *artifact* of preservation. The first recorded Cambrian outcrops preserved *only* species with hard, *shelly* parts, such as *trilobites, brachiopods, mollusks*, and *echinoderms*. When the *Burgess Shale, Chengjiang*, and other similar Cambrian *Konservat-Lagervat-Lagerstatten* (remarkably preserved fossil deposits) were discovered, large numbers of more delicate Cambrian arthropod species (that typically do not preserve at all) were revealed as *contemporaries* of trilobites, rather *than* their descendents. Trilobites are now, nevertheless, considered *relatively advanced* among the Paleozoic arthropods, and the search for the *first* arthropods and the *ancestors of trilobites* is *pushed* back into the Precambrian.

Status of "Trilobitomorpha"

In the 1959 *Trilobite* Treatse, many of the *Burgess Shale* arthropods were included in the arthropod *Subphytum Trilobitomorpha*. Today that subphytum is no longer considered valid – it served as a *convenient* unit for various Paleozoic *taxa* with *similarities* in limb structure, but has *fallen* aside for more *rigorous* analysis. One prevalent recent classification recognizes the *Trilobita* as a *class* of *arthropods* sitting *comfortably* within the *Superclass Archnomorpha* (an expanded concept based on the Chelicerata), alongside the Superclass Crustacumorpha (based on an expansion of the Crustacca), together comprising the *Subphytum Schizorramia* (Arthropods bearing *biramous* limbs) contrasting with the *Subphytum Ateloeerata* (insects, myriapods, and allies). This classification diagram is shown.

While there is certainly a great deal of *diversity* of forms among the *arthropods* and *Near-arthropods* (such as the large predatory *protarthropodan Anomalocaris*), it seems reasonable to assume that shared *common ancestry* in the *Pre-Cambrian* was the basis for the *radiations* of the early *Cambrian*, such as seen at the Burgess Shale (Canada)

Chengjiang (China) and Sirius Passet (Greenland). This shared ancestry and close relationship despite *seemingly* great *divergence* of *form* is seen in the very *similar molecular biology* of *modern crustaceans* and *insects*, and suggests that the *Subphylum* and *Superclass* designations for the *Atrthropoda* may be superfluous (especially that dividing the *Atelobita* from the *Schizoramia*)

Archnomorpha and Crustaeomorpha

Among the *diversity* of Paleozoic arthropods, *two large groups* have emerged in recent phylogentic analysis: a group of *crustacean-like arthropods* referred to as *Crustaceoimorpha*, and a group referred to as *Arachnomorpha*. The diversity within the **Arachnomorpha** is dominated by *Trilobita* and *Chelcerata*, each forming major *clades*.

Archnomorpha (Stormer **1944**) is equivalent to a grouping called *Archnata* (Lauterbach **1983**), which is defined as *an inclusive grouping of non-crustacean arthropods: a clade stemming from the ancestor of Trilobita and Chelicerata*. The Arachnomorpha cladogram and its two major clades (Sensu, Cotton, and Braddy **2004**).

Archnomorph diversity allows attempts to elucidate the *relationships* between t*rilobites* and other Paleozoic archnomorphs. It is relatively easy to *exclude* a large subset of archnomorphs that fall into a *Chelicerate Clade*," which can generally be c*haracterized* as *archnomorphs*, bearing a post-anal telson/tail, and a pair of limbs per segment (among other sympomorphies). The *Burgess Shale* artheopods *Sidneyia* and *Yohoia* are two good example members of this *Chelicrate Clade*.

Features of the "trilobite Clade"

Most of the archnomorph trilobite clade share features *in common* with trilobites, such as a hypostome, **3-4 pairs** of post-antennular legs under the cephalon, *similar* limb structure, etc. Such similarities have led to proposals to *include* some trilobite-like archnomorph in the class Trilobita (e.g., the *Naraoiidae*). Some recent cladistic analysis of arachnomorph taxa, however, has elevated Helmetiidae as a clade *closely* aligned with the trilobites (see figure), which would create a parapphyletic situation if the naraoiids are a*lso* considered trilobites. It seems *more* reasonable to recognize the *similarities* of members of the "Trilobite Clade" of the Arachnomorpha, while *also* recognizing their distinctiveness.

INTRODUCING THE EVOLUTION – FROM REPTILES TO MAMMALS

Synpsids

The most obvious *distinction* between mammals and reptiles are the fact that mammals have *hair* or *fur*, and *mammary* glands which they use to *nourish* their young. These features do not *fossilize*, and no known mammals have *left* hair or fur *impressions* in the rock surrounding their fossils. Fortunately, however, there are also a number of skeletal d*ifferences* between reptiles and mammals. For one, reptiles have a mouth filled with several teeth which are more or less *uniform* in size and shape; they vary slightly in size, but they all have the same basic *cone-shaped* form. By contrast, mammals tend to have teeth which *vary* greatly in size and shape; everything from flat, *multi-cusped* molar teeth to the sharp *cone-shaped* canines. In reptiles, the lower jaw is comprised of *several different bones*, which hinge on he quadrate bone of the skull and the angular one of the jaw. In mammals, however, the lower jaw is comprised of only *one bone* – the *dentary*, which hinges at the *quadrate* of the skull. In mammals, there *three bones* in the *middle ear*, the *malleus*, stapes (also known as the *hammer, anvil* and *stirrup*). In reptiles, there is only *one bone* – the *stapes*. The reptilian skull is attached to the *spine* by a *single* point of contact, the *occipital condyle*. In mammals, the *occipital condyle* is "double-faced." The classic reptilian skull also has a *small hole*, or "third eye" through which the pineal body extends – a trait *not* found in any known mammals. This brings us to the **synapsid** reptiles. Like mammals, they have a *single*, lower temporal fenestra. Already this makes them *more akin* to *mammals* than other *reptiles*, albeit though, still with a very *reptilian* body; legs *sprawled* out, long *whip-like* tail, basically *conical-uniformed* teeth, etc. One group of *synapsid* reptiles in particular, the **therapsids**, seem to *break* these rules and are *adorned* with very mammilian characteristics; in the more *advanced* forms, many of the bones *absent* in mammalians were already being *reduced* to near *extinction*, and the "third eye" *so small* it might as well have been *absent*.

Likewise, in many of these forms, there was a sharp *contrast* between the different types of *teeth*; incisors, broadly *crowded* cheek teeth with accessory *cusps*; the *occipital condyle* also because "double-faced" in many of these reptiles.

Colbert and *Morales* (**1991-p.127**) describe the *transitional* nature of the **tritylodonts** in particular: "In many respects, the *tritylodont skull* was very *mammlian* in its features.

Certainly, because of the advanced nature of the *zygomatic* arches, the *secondary* palate and the specialized *teeth*, these animals had feeding habits that were *close* to those of some mammals . . . . Yes, in spite of these advances, the *tritylodonts* still *retained* the reptilian joint between the *quadrate bone* of the *skull*, along with the *articular* bone of the l*ower* jaw. It is true that these bones were very much *reduced*, so that the *squamosal* bone of the skull and the *dentary* bone of the *lower* jaw (the two bones involved in the mammalian jaw articulation) were on the point of *touching* one another."

Creationists (or,"*Intelligent Designers*") are, however, quick to try and *discredit* the *Fossil* evidence for *mammalian evolution* from *reptiles*. *Gish* (**1978. p. 80**) claims that:

"The *two* most distinguishable osteological *differences* between *reptiles* and *mammals*, have *never* been *bridged*, however, by a *transitional* series. All mammals, *living* or f*ossil*, have a *single* bone, the *dentition*, on each side of the *lower* jaw, and all mammals, *living* or *fossil*, have *three* auditory *ossicles* or *ear bones*, the *malleus, incus* and *stapes*. In some fossil reptiles the *number* and *size* of the *lower* jaw-bones were r*educed* compared to *living* reptiles. Every reptile, living or fossil, however, has/had at least *four* bones in the lower jaw and only one auditory *ossicle*, the *stapes* . . . . There are/were no transitional forms showing, for instance, *three* or *two* jawbones, or *two* ear bones. No one has explained yet, for that matter, *how* the *transitional* form would have managed to *chew* while his *jaw* was being *unhinged* and *rearticulated*, or how it would *hear* while dragging *two* of its *bones* up into its ear."

This *transitional* series is not as *mythical* as Gish is trying to purport. Indeed, the transition can *clearly* be seen in *Triassic therapsids* – though obviously occurring much *differently* than Gish has *described*! (1995) writes:

"The *reptiles*, as we have *noted*, have *one* bone in the *middle ear* and *several* bones in the *lower* jaw, while *mammals* have *three* bones in the *middle ear* and only *one* bone in the *lower* jaw. On the other hand, however, the *jaw joints* in the reptile are *formed* from d*ifferent* bones than they are in the mammalian *skull*. It is apparent that during the evolutionary *transition* from *reptile* to *mammal*, the *jaw joints* must have shifted from *one* bone to *another*, freeing up the *rest* of these bones to *form* the *auditory ossicles* in the mammalian *middle ear* – The fact *is* that, in *most* modern reptiles, the jawbones **in question actually function in transmitting sound waves in the inner ear**, so the *transformation* postulated above **is not a functional change**, merely an i*mprovement* (the *term* between a **new function** and a **modification** in describing any animal in *evolutionary transition* is often *misdirected* – author's note) in a *function* that these bones already *had*. As *Arthur N. Strahter* puts it, "A *transitional* form must have had two joints in operation simultaneously (as in the *modern* rattlesnake), and this phase was followed by a fusion of the lower joint."

(*Strahter* 1987. p. 414) . . . . Not only is this explanation not 'merely wishful conjecture', but it can be clearly seen in a remarkable series of fossils from the *Triassic therasids*. The *earliest therapsids* show the *typical* reptilian type of *jaw* joint, with the *articular* bone in the jaw firmly *attached* to the *quadrate* bone in the skull. In *later* fossils from the *same* group the *quadrate-articular* bones, however, have become *smaller*, and the *dentition* and *squamosal* bones have become *larger* and have m*oved* closer together. This trend reaches its *apex* in a group of *therapsids* known as **cynodonts**, of which the genus *Probainognathus* is a *representative*. Probainognathus possessed characteristics of both *reptile* and *mammal,* and this *transitional* aspect was s*hown* most clearly by the fact that it had **two** jaw joints – one *reptilian*, and one m*ammalian*."

"*Probainognathus*, a small cynodont reptile from the *Triassic sediments* of Argentina, exhibited characteristics in the *skull* and *jaws* far *advanced* towards the mammalians.

Thus, it had teeth differentiated into *incisors*, a *canine, post canines*, a *double occipital condyle* and a well-developed *secondary palate*, all features typical of *all mammals*. But most *significant* of *all*, the *articulation* between the *skull* and the *lower jaw* was on the very *threshold* between the *reptilian* and *mammalian* condition. The *two* b*ones* forming the articulation between skull *mandible* in the *reptiles*, the *quadrate* and the *articular* respectively, were *still* present but were *very small*, and *loosely* joined to the *bones* that constituted the mammalian *joint* . . . . In *Probainognathus* there was, therefore, a *double* articulation between *skull* and *jaw*, and of particular interest, the quadrate bone, so *small* and so *loosely* joined to the *squamosal*, was intimately articulated with the *stapes* bone of the *middle ear*. It quite obviously was well on its way towards being the *incus* bone of the three-bone complex that characterizes the mammalian *middle ear.*

Next in the reptile-to-mammal transition sequence are the *cynodonts*. The **Cynognathus**, a *classic* example of the *cynodont* reptiles. When faced with a specimen such as this, one is forced of course, to wonder if it can be truly called "reptile." The s*kull* appears *basically* mammalian, the same goes for the *hip structure*, but with very distinct *similarities* to reptiles as well. The *grastral ribs* and *vertebrae* seem to be forming a *primitive* breast-bone (*sternum*) "floating" ribs have been *reduced* to *almost* nothing, and are completely *absent* in *mammals*, yet very *large* in *reptiles*. The animal isn't *quite* a mammal, but it isn't *quite* a reptile *either*, a creature, more or less, *split* down the middle – being *half* reptile and *half* mammal.

Diapsid, Archosaur – True Reptile

Euparkeria

Late Permian

Synapsid, Theropsid – "Reptile"?

Cynognathus

Early Triassic

Synapsid, Thrapsid – True Mammal

Canis (specifically a Grey Wolf)

Mid Territarry/Late Quarternary

Oligokyphus

Shortly after the time of the *Cynodonts*, a gap is found in the Mid-Triassic (*239-208 million years ago*). Until recently, there were no known *therapsid* fossils occupying this area of geological time. *Adelobaileus cromptoni*, a fairly "new" species, however, has been discovered; and at age **225 million years**, lands squarely in the *middle* of the Triassic gap. Though only a *skull* was found, some cranial features of *Adelobaileous*, such as the incipient *promontorium* housing the *cochlea*, represent an *intermediate* stage of the character *transformation* from *non-mammalian* cynodonts to Triassic *mammals*".

The *proto-mammal* **Adelobaileus** is thought to be either (1) the *common* ancestor of mammals, or (2) a very *close* relative of that *common ancestor* (Hunt, 1947).

The next *proto-mammal* (**Sinoconodon**) appeared 208 million years ago. Its cheek- teeth were *permanent*, as in *modern* mammals, nevertheless, as in reptiles, the *other* teeth were still replaced *several times* during its *lifetime*. The mammalian-joint of the *jaw* was "*stronger*, with *large dentary condyle* fitting into a distinct *fossa* on the *squamosal* . . . . This *final refinement* of the *joint* automatically made this animal a *true* 'mammal'. . . . The reptilian *jaw joint* was still present, though tiny." (Hunt, 1997?) The *rear* of the braincase had also *expanded* with the eye socket *fully mammalian*. Hunt (1997) describes a group of *proto-mammals* which appeared roughly *3 million years* after *Sinoconodon*:

"*Eozostrodon, Morgonucodon, Haldanondon*, (early Jurassic, **205 Million years ago**) A group of early *proto-mammals* called "**morganucodonts**". The teeth were *truly mammalian*, with *cheek* teeth finally *differentiated* into simple premolars and more complex molars, the teeth replaced just *once*. There were Triangular-cusped molars, a r*eversal* of the previous trend towards *reduced* incisors, with the lower *ones* increased to *four*. The *tiny* remnant of the reptilian *jaw-joint*, once thought to be ancestral to monotremes *only*, are now, however, taught to be ancestral of *all* three groups of m*odern* mammals – *monnotremes, marsupials*, and *placentals*." The mammals continued to evolve and with several other links known from the mesozoiz era. "By the late Cretaceous, the three groups of modern mammals were in place: *monotremes, marsupials*, and *placentals*.

Dimetrodon Grandis
Dimetrodon: *ancestor of all Mammals* belonged to the family called **Pelycosaurs**, had both m*ammal*and *reptile* *c*haracteristics.

Dimedrodon Grandis

It *preceded* the *earliest* dinosaurs by more than *40 million years*, but *physically,* however, It *looked* a lot like one. *Dimetrodon* is often referred to as a *mammal-like* reptile, based on characteristics of the *skull* and *dentition*. It was a *dominant* carnivore, the *largest* of the *Permian Period,* a ferocious, predacious reptile that was on *top* of the *food chain* during the *Early Permian*.

This *pelycosaur* possessed a spectacular *sail* on its back, *supported* by *long, bony spines*, each of which *grew* out of a *separate* spinal vertebrae. The sail was probably an e*arly* experiment in *controlling* body temperature. It is believed that the sail *absorbed* the h*eat* of the sun to *warm* its blood. It warmed up *early* after sunrise and cooled off more efficiently during the *heat* of the day, which also may have been used for *mating* and d*ominance* rituals together with making it look *larger* than it was to predators.

Dimetrodon had a *large* skull with *two* types of *teeth* – sharp canines and shearing teeth. *Dimetrodon* was about **11.5 feet (3.5 m)** long and weighed roughly **550 pounds (25 kg)**. Having to walk on all four *side-sprawling* legs, it probably lumbered along at quite s*low* gait.

INTRODUCING ARCHOSAURS

The Great Archosaur Lineage: Crocodiles, dinosaurs, ptersaurs, along with many other beasties!

Archosaurs (Greek for '*ruling* Lizards') were a group of *diapsid* reptiles represented by modern *birds* and *crocodiles* – a group also includes extinct *non-avian* dinosaurs, p*terosaurs* along with *relatives* of crocodiles.

There is some *debate* about when *archosaurs* first introduced *themselves*. Those classify the *Permian* reptiles *Archosaursrossicus* and/or *Protorosaurs speneri* as *true* archosaurs maintain that *archosaurs* first appeared in the *Late Permian*. Those who classify both *Archosaurs rossicus* and *Protorosaurs speneri* as Archosaursi*formes* (not true *archosaurs* but *closely* related) maintain that *archosaurs* first evolved from *Archosauriformes* ancestors during the *Olenckian – Early Triassic*.

Distinguishing Characteristics

The simplist and most widely-agreed synnapomrphies of archosaur are:

- Teeth *set in sockets*, which makes them *less* likely to be torn loose during feeding. This feature is responsible for the name "*theco[donts]*" ("socket teeth"), which paleontologists used to apply to all or most *archosaurs*. Some *archosaurs*, such as b*irds*, are secondary[ily] toothless.

- *Antorbital fenestrae* (openings in the skull in *front* of the eyes but *behind* the nostrils), which *reduced* the weight of the skull, a *useful* feature since most early *archosaurs* had long, heavy skulls, rather like those of modern crocodilians. The *preorbital* f*enestrae* (sometimes called *anteorbital fenestrae*) are often larger than the *orbits (eye sockets)*

- *Mandibular fenestrae* (small openings in the jaw bones), which may have *reduced t*he *weight* of the jaw slightly.

- A fourth *trochanter* (ridge for attaching muscles) on the femur. This seemingly insignificant detail may also have been connected with the ability of the *archosaurs* or their immediate ancestors to survive the catastrophic *Permian-Triassic* extinction event.

Archosaur Takeover in the Triassic

The *Synasida* (informally known as "mammal-like reptiles") were the dominant land animals throughout the *Permian*, but most perished in the *Permian-Triassic* extinction event – *Lystrosaurus* (a *herbivore* mammal-like *reptile*) was the *only* large land animal to survive the event, becoming the most *prevalent* land animal on the *planet* at that time period.

The *archosaurs* didn't waist much time in taking over and becoming the *dominant* vertebrates in the *Early Triassic*. The two most commonly-suggested explanations for this:

- *Archosaurs* made more rapid progress than *mammal-like* reptiles towards *erect* limbs, giving them greater *stamina*, by avoiding *Carrier's constraint*. This, however, is questionable, since *Archosaurs* became dominant while still *ambulating* with s*prawling*, or *semi-erect* limbs, similar to those of *Lystrosourus,* along with other m*ammal-like* reptiles.

- The *Early Triassic* was predominantly *arid*, because *most* of the earth's landscape was concentrated in the *super-continent* **Pangaea**. *Archosaurs* were probably better at conserving *water* than the *mammal-like* reptiles, because:

- Modern *diapsids* (Lizards, snakes, crocodiles, birds, etc.) excrete *uric acid*, which can be excreted as a *paste*. It is not *far-fetched* to say that *archosaurs* (*diapsids,* and the ancestors of *crocodilians, dinosaurs and bird, etc.*) also excreted *uric acid*, and therefore were good at conserving *water*. The *aglandular* (glandless) skin of *diapsids* would have also helped to conserve *water*.

- Modern mammals excrete *urea*, which requires a lot of water to keep it dissolved. Their skins also contain *glands,* which also *lose* water. Assuming that *mammal-like* reptiles had similar features, e.g., as argued in *Palaeos*, they were at a *disadvantage* in a primarily *arid* world. This same well-respected *site* points out that "for much of Australia's *Plio-Pleistocene* history, where conditions were probably similar, the largest terrestrial predators were *not mammals* but gigantic *varanid lizards* ("Megalania") and *land crocs*."

It has also been suggested that the Triassic was low on oxygen and archosaurs had a more advanced respiratory system.

Main Types of Archosaurs

Since the **1970s** scientists have classified *archosaurs* on the basis of their *ankles*. The earliest *archosaurs* had "*primitive mesotarsal*" ankles – the *calcaneum* were *fixed* to the *tibia* by sutures and joint bent about the contact between those bones and the foot.

The *Crurotarsi* appeared early in the *Triassic*. In their ankles the *astragalus* was joined to the *tibia* by the suture and the joint *rotated* round a peg on the *astragalus* which fitted into a *socket* in the *calcaneum*. Early "crumtarsans" still walked with *sprawling limbs*, but some time later "cruurotarsans" developed *fully erect* limbs (most notably the *Rauisuchia*).

Modern *crocodiles* are also "crurotarsans" that walk with their limbs *sprawled*, or depending on how much of a hurry they are in. *Euparkeria* and the *Ornithosuchidae* had "*reversed crurotarsal*" ankles, with a peg on the *calcaneum* and socket on the *astragalus*. The *earliest* fossils of *Ornithodira* ("bird necks") appear in the *Carnian Age* of the *Late Triassic*, but it is hard to see how they could have evolved from the "*crurotarsan*" – They possibly evolved much earlier, or perhaps they evolved from the *last* of the "primitive mesotarsal" *archosaurs*. *Ornithodires* "advanced mesotarsal," in which their a*nkles* had a very large *astragalus* and very small limbs, but provided more stability when the plane, like a simple hinge. This arrangement was suitable only for animals with erect limbs, but provided more *stability* when the animals were *running*. The o*rnothodires* differed from other *archosaurs* in other ways – they were *lightly-built* and usually *small*, their necks were *long* and had an *S-shaped curve*, their *skulls* were much more *lightly built*, and many *ornothodires* were *completely bipedal*. The *archosaur*s fourth *trochanter* on the *femur* may have made it easier for *ornothodires* to become b*ipedal*. Because it provided more *leverage* for the *thigh muscles*. In the *Late Triassic* the o**rnothodires** diversified to produce **pterosaurs** and **dinosaurs**.

Hip Joints and Locomotion

Like the early tetropods, early archosaurs had a sprawling gait because:

* Their hip sockets faced *sideways.*
* The knobs at the tops of their femurs were *in line* with the femur.

In the early to *Mid-Triassic*, some *archosaur* groups developed *hip joints*, which allowed (or better put, *required*) a more *erect gait*. This gave them *greater stamina*, because it avoided *Carrier's Constraint*, i.e., they could *run* and *breath* more easily at the same time.

There were *two* main types of joints which allowed erect legs:

* The *hip sockets* faced *sideways,* but the *knobs* on the *femurs* were at *right angles* to the *rest* of the *femur*, which therefore pointed *downwards. Dinosaurs* evolved from *archosaurs* with *this* hip arrangement:
* The hip sockets faced *downwards* and the *knobs* on the *femurs* were *in line* with the femur. This "pillar-erect" arrangement appears to have evolved more *independently* in various *archosaur* linsages – it was, for example, common in *Rauisuchia* and also appeared in some *actosaurs.*

Extinction and Survival
Crocodilians, pterosaurs, and *champsosaurs* survived the *Triassic-Jurassic* extinction event about *195 million years ago*, but other *archosaurs types* became extinct.

Non-avian *dinosaurs* and *pterosaurs* perished in the *Cretaceous-Tertiary* extinction event, and it is generally agreed that *birds* (last surviving *dinosaur group*) survived. *Birds* are descendents of *archosaurs*, and are therefore, still *classified* as a*rchosaurs* themselves under *phylogenetic taxonomy.*

Champsosaurs became extinct in the *Early Miocene.*

Crocodilians (which include all modern *crocodiles, alligators*, and *gharials*) and *birds flourish* today, and it is generally conceded that *birds* comprise the *most species* of all t*errestrial vertebrates.*

Archosaur Lifestyle
Most were large predators, but members of various lines diversified into other niches:

* *Actosaurs* were *herbivores* and some developed spectacular *armor.*

* A *few* crocodiles were *herbivores*, e.g., *Simosuchus, Phyllodonosuchus.*

* The large crocodilian *Stomatosuchus* may have been a *filter feeder.*

* *Auropdomorphs* and *ornithichian* dinosaurs were herbivores with *diverse* adaptations for feeding biomechanics.

Land, Water and Air:
Archosaurs are mainly portrayed as land animals, but:

* The crocodilians *dominated* the *rivers* and *swamps* and even *invaded* the sea areas, e.g., Metiorhynchidae and Dyrosauridae. The *Metriorhyidae* were somewhat d*olphin-like* mammals, along with *paddle-like* forelimbs, a tail fluke and smooth, unarmored skin.

* Two clades of *ornithodirians*, the *pterosaurs* and the *birds*, dominated the air becoming adapted to a *valant* (violent) lifestyle.

Matabolism
The *metabolism* of *archosaurs* is still a controversial topic. They unquestionably *evolved* from *cold-blooded* ancestors, and the surviving non-dinosaurian *archosaurs*, were cold- blooded. But *crocodilians,* however, posses *some* features which are normally associate with a *warm-blooded* metabolism because it improved the animal's oxygen supply.

* *4-chambered hearts. Mammals* and *birds* have 4-chambered hearts. Non-crocodilian reptiles have 3-chambered hearts, which are less efficient because they allow oxygenated and de-oxygenated blood to mix and therefore send de-oxygenated blood out of the body instead of the lungs. Modern crocodiles' hearts are *4-chambered,* but are smaller relative to the body size and run at lower pressure than those of mammals and birds. They (crocodiles) also have a *bypass,* which makes them functionally *3- chambered* when under water, conserving oxygen.

* a *secondary* palate, which allows the animal to *eat* and *breathe* at the same time. This is different from the l*ung-pumping* mechanisms of mammals and birds but similar to what some researchers claim to have found in some dinosaurs.

* a hepatic piston mechanism for pumping the lungs. This is different from the lung- pumping mechanisms of mammals and birds but similar to what some researchers claim to have found in some dinosaurs.

So, why did *natural selection* favor the development of these features, which though, very important for active warm-blooded creatures, are of little apparent use to cold- blooded aquatic ambush predators that spend the majority of their time floating in water or lying on river banks?

Some experts believe that crocodilians were originally *active*, *warm-blooded* predators and that their *archosaur* ancestors were *warm-blooded*. Developmental studies indicate that crocodilian embryos developed fully *4-chambered* hearts *first* and then developed the modification which make their hearts function as *3-chambered* under water. Using the principle that *ontogeny recapitulation phylogeny*, the researchers concluded that the original crocodilians had fully 4-chambered hearts and were therefore *warm-blooded* and that later crocodilians developed the *bypass* as they *reverted* to being *cold-blooded* aquatic predators.

If the original crocodilians were warm-blooded and other Triassic *archosaurs* were also warm-blooded, this would help resolve some evolutionary puzzles:

* The *earliest* crocodilians, e.g., *Terrestrisuchus*, were *slim, leggy* terrestrial predators whose build suggests a fairly *active* lifestyle, which requires a fairly *fast metabolism.*

  And some other "crurotarsan" *archosaurs* appear to have had *erect* limbs, while those of *rauisuchians* were very poorly adapted for any other posture. Erect limbs are a*dvantageous* for active animals because they avoid *Carrier's constraint*, but d*isadvantageous* for more sluggish animals because they increase the energy costs of standing up and lying down.

* If early *archosaurs* were *completely cold-blooded* and (as seems most likely) dinosaurs were at least *fairly warm-blooded*, dinosaurs would have had to evolve warm-blooded metabolism in less than half the time it took for mammal-like reptiles to do the same.

Postsuchus

This *continues* the **Tetrapod** trend of the *reduction* of the skull bones by the *fusion* of multiple bones and the opening of *fenestrae* in the skull. This helps to *lighten* the skull, provides more room for *muscles* and other *tissues*, allowing more skull *flexibility* (kinesis) when eating. Other typical archosaurian characteristics include another opening in the lower jaw (the manibular fenestra), a *high narrow* skull with a *pointed* snout, teeth set in sockets (called the *codont* tooth implantation), and a *modified* ankle joint.

The ancestral *archosaurs* probably originated some **250 million years ago**, in the late **Permian Period,** around **286-245 million years ago**). Their descendents (such as dinosaurs) dominated the realm of the terrestrial vertebrates for a majority of the **Mesozoic Era** (**245-66 million years ago**), for *roughly* **140 million years**. Today, only the *birds* and *crocodiles*, along with a scattering of nondescript *lizards* exist to provide a glimpse into the past glory of archosaurs.

On dry land the *diapsids* originally *ambulated* in the same *sprawling* manner as l*izards*, with *extended* limbs *spreading out* from the *body*, moving them *horizontally* in w*ide arches*, with their body *waving* from *side* to *side*.

But when they *swam,* however, they would hold their limbs *close* to their body so as to *minimize* the water *resistance*, and when *leaving* the water to *occupy* the banks, they eventually evolved *thighs* and *knees* in *close* proximity to the body *flanks*. They *walked* by *swinging* the *lower* parts of their *hind* legs *vertically*, with their feet *directly under* the body. They additionally evolved a *narrow-track gait*, which enabled them to *walk* on their *hind* legs alone, balancing the *forepart* of their body by their heavy muscular tail.

This gait proved *advantageous* as it required much *less* muscular effort than the s*prawling* style of walking of their *diapsid forebears*, resulting in a *key* change in the evolution of the *original* breast-pelvic skeleton. Since the *forelegs* were relieved from the b*urden* of the *body-weight,* the *breast-skeleton*, apt for the strong muscular four- legged *sprawling* reptile, was *reduced* in the dinosaur design model, except in *birds* where it remains a *necessary* support for its on-going flight.

The *pelvis* consisted of the *iliac* bone, which was *attached* to the *spine* by a few vertebrae, the *pubic* bone, connected to the *iliac* bone, faced *forwards*, while the i*scium* bone faced *backwards.*

The bipedal, narrow-tracked gait, in which the pelvis *alone* should carry the *entire* w*eight* of the body, needed however, *more* vertebrae to *attach* the *spine* to the *iliac* bone, and consequently, was *elongated. Bipedal gait* also caused the muscles (which in the q*uadrupedal diapsids* had pulled sideways-extended thighs *downward*) to pull forward- extended thighs *downward*, as the *hind* legs carried the *entire* body-weight. As these muscles were *attached* to the *pubic* bones, they were elongated and bent *downward* in order to *reinforce* these muscles. This *pelvis* structure with the *long* ilium and the *long*, d*ownward* faced *pubis* is also characteristic of the bipedal *archosaurs* and *saurischian* dinosaurs.

Terrestrisuchus

As a consequence of the bipedal, narrow-tracked gait, the *archosaurs* could develop l*ong* hind legs. Later on, some of them also lifted their heels *off* the ground to become i*gitigrades*, which *changed* the ankle-joint into a simple hinge.

The swimming diapsids

Other *diapsid* reptiles survived by staying *in* the water, *living* on fishes and *swimming* by w*aving* their tails. As an adaptation this mode of life, their *jaws* were *elongated* so as to hold their nostrils *above* the *water*, which resulted in the creation of their *antorital* opening. They also developed the solid *thecodont* dentition to *catch* and *hold* fish m*ore* securely. Other *diapsid* reptiles survived by staying *in* the water, *living* on fishes and *swimming* by *waving* their tails. As an adaptation this mode of life, their *jaws* were e*longated* so as to hold their nostrils *above* the *water*, which resulted in the creation of their *antorital* opening. They also developed the solid *the codont* dentition to *catch* and h*old* fish *more* securely.

INTRODUCING THE DINOSAUR

Dinosaurs, *one* of the *most successful* groups of animals (in terms of *longevity*) that have e*ver lived*, evolved into *many* diverse sizes and shapes, with many equally diverse modes of living. The term "*Dinosauria*"was *created* by *Sir Richard Owen in* **1842** to *describe* these "fearfully great reptiles," specifically *Magalosaurus, Iguanodon*, and *Hylaeosaurus*, the *only* three dinosaurs *known* at the time. The creatures that we normally think of as dinosaurs lived during the *Mesozoic Era*, from late in the *Triassic Period* (about **225 million years ago**) until the end of the Cretaceous (about **65 million years ago**, a time-span of about **160 million year***s*). But we are now aware that they actually l*ive on* to this day as the **birds**.

Some things to keep in mind about dinosaurs:

Not everything big and dead is a dinosaur

All too often, books *written* (or movies) *made* for popular audiences include animals such as *mammoths, mastodons, pterosaurs, plesiosaurs, ichthyosaurs*, and the *sail-backed Dimetrodon*. Dinosaurs are a specific subgroup of the **archosaurs**, a group that also includes *crocodiles, pterosaurs*, and *birds*, although *pterosaurs* are closely related, they are *not* dinosaurs in the *true* sense of the word. Even more *distantly* related to the dinosaurs are *marine* reptiles, which include the *plesiosaurs* and *ichthyosaurs*.

Hylaeosaurus

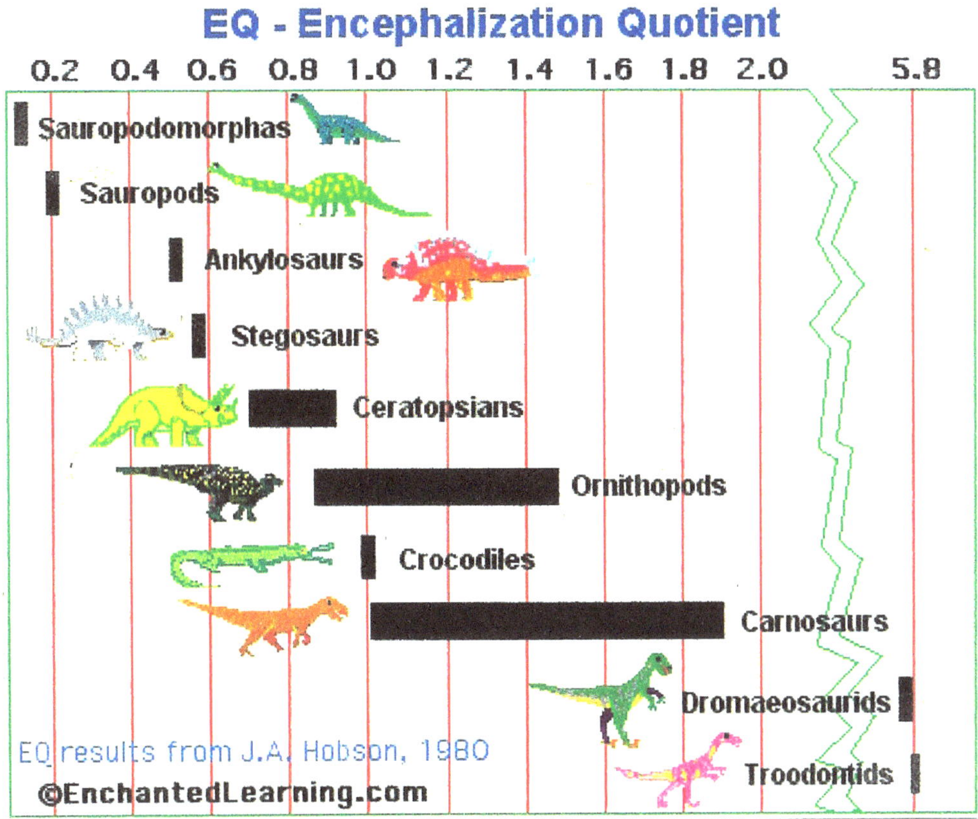

Mammoths and *mastodons* are mammals that did *not* appear *on* the scene until many millions of years *later*, following the end of the *Cretaceous Period*. *Dimetrodon* is neither a *reptile* nor a *mammal*, but a *basal synapsid*, i.e., an *early* relative of the ancestors of mammals.

Not all dinosaurs lived at the same time

Different dinosaurs lived at *different* times. Despite the *portrayals* in movies like *King Kong* and *Jurassic Park*, no *Stegosaurus* ever saw a *Tyrannosaurus*, because *Tyrannosaurus* didn't appear on the scene until *80 or million years* following the extinction of *Stegosaurs*. The same goes for *Apatosaurus* ("*Brontosaurus*") – Its bones were *long before* fossilized by the time of *T-rex*.

Dinosaurs have not gone extinct

Technically, based on the features of the skeleton, most people studying dinosaurs consider *birds* as a continuation of dinosaurs. So rather than refer to "*dinosaurs*" and "*birds*" as discrete, separate groups, it is best to refer to the traditional *extinct* animals as "*non-avian*" dinosaurs, and *birds,* as *birds*, or "*avian*" dinosaurs, as some would say, "*pro-avian*" dinosaurs. *No* creature that has left living descendents behind can be regarded as "extinct," as in the case of the dinosaur that has left behind such living descendents as the *cockatoos, cassowaries*, to mention just a few of the bird types – as also is the case with vertebrates, even though their Cambrian *ancestors* have long since gone extinct.

How the dinosaurs originated and disappeared

Ancient Connections

The dinosaurs originated from *diapsid* reptiles, which survived the *mass-extinction* at the *end* of the *Permian*, as *swimming archosaurs*. From those emerged *bipedal* **archosaurs**, among the other two groups of bipedal herbivores, the short-necked ancestors of the *ornitschian* dinosaurs, and long-necked ancestors of the *saurrisehian* *sauropods*. The short-necked meat-eaters emerged later, ancestors of the *caraosaur* dinosaurs. The various groups of bird-like *coeurosaurs*, branched from a lineage of *small* bipedal, *tree-climbing* bird ancestors.

Preposition

This article on the *origin* and *disappearance* of the dinosaurs is based on the *supposition*, forwarded and argued for on *The Origin of Flapping Flight of Birds*, that the dinosaurs are *offshoots* from a lineage which originated from a group of *diapsid* Permian r*eptiles* leading to the *birds*.

The *source* group of reptiles, which were very much like the recent **Sphenodon, w**hich *evolved* into *birds*, the only group of recent bipedal diapsids, through continuous stages, each stage set for behavioral, physiological and anatomical qualities to better a*dapt* and *breed* in the actual environments than the former groups. The characters typical for dinosaurs, evolved gradually with the *Jurassic Period* pre-bird *Archaeopteryx*: elongated *jaws* with *thecodont* teeth and *antorbital* opening and the bipedal, narrow- tracked gait. In the course of time *many* offspring of these various groups of dinosaurs b*ranched off* from this lineage.

Considering the relationship of bird evolution to dinosaurs, along with the mutual relationship between the various groups of dinosaurs among one another, one is presented with the question of, exactly what specifically defines a dinosaur? Much *time* and *space* could be spent on that. To *avoid* that, I prefer to call the entire group *archosaurs*, and only those in the community that are called *dinosaurs*, i.e., the *carnosaurs*, the *pro- sauropods*, and the four groups of *ornitischians,* as a generic term for all dinosaurs.

Hatteria (order Sphenodon)

The *skull* of the dinosaur is generally characterized by *two* temporal openings behind the eye. As this is also a characteristic trait of a group of Permian diapsid reptiles, with little difference from the non-specialized advanced reptile *Sphenodon*, which were supposed to have been the *stem group* of the dinosaurs. The diapsid *skull* was also a characteristic trait of *crocodilians*, and *Archaeopteryx*, and like the dinosaurs these also have the particular opening in *front* of the eye, the *antorbital* opening, and they have a s*ingle* row of teeth in caves of the *jaw,* and *thecodont* dentition. These traits are a general characteristic of the *archosaurs* to which also include the *quadrupted* reptiles from the Triassic, the *phytosaurs* and the *aetosaurs.*

The *narrow bipedal* gait is a general characteristic of *archosaurs*, that also generally describes the traits of birds, among other groups of dinosaurs.

The dinosaurs are *distinctive* in two main groups, the *onitischians*, and the s*aurischians*, characterized by a particular structure of *pelvis* and *dentitition.*

The antororbital opening, the theodont teeth, the narrow track way, bipedal gait, the different pelvic structure, as well as all the other characteristics of the various groups of dinosaurs, together with a different lifestyle, sets them apart from their *reptile* ancestors, ushering in the ***dinosaur*** as a ***new origin of species***. Thus, the ancestors of the s*aurischian* dinosaurs must have lived quite differently from the ancestors of the o*rnitischians.* The ancestors of the four-footed *theropods*, the forefathers of the heavy, short-necked *carnosaurs* undoubtedly had a different lifestyle than those of the light, long necked *coelurosaurs*. This applies also to the ancestors of the two-legged *ornithopods*, with three toes at each foot.

The following is sketchy proposal on what caused the differences in the general structure of the various groups of dinosaurs, their relationships with one another, the consequences, and finally – their extinction!

History

The dinosaurs *emerged* during the Triassic Period. In the previous Permian Period a m*ultitude* of *vertebrates* had arisen, *among which* were **our mammalian ancestors**, along with some lizard-like *diapsid* reptiles, not much different than *Sphenodon.*

During the Permian the climate was *warm* and *dry*, but by the end of the period it became *so* hot and dry, that it turned *vast* areas of the earth into *deserts*, causing the e*xtinction* of *many* species, with just a *few* surviving to usher in the *Triassic Period.*

Some the *few* remaining survivors included the *diapsid* reptiles, which were able to seek shelter from the sweltering *heat* of the *sun* under the *cooling comfort* of the *rocks*, venturing out *only* to search of food, and *only* at *dawn* and *after sunset*. Dinosaurs, lizards and snakes arose from these same *diapsid* reptiles.

In the early Triassic, when the climate became *less* dry and *plants* began to *spread*, another *group* of the lizard-like *diapsids* evolved into the plant-eating **rhynchosaurs**, from which *many* species arose. They were *tall* and *clumsy* reptiles, but disappeared b*efore* the end of Mesozoic period.

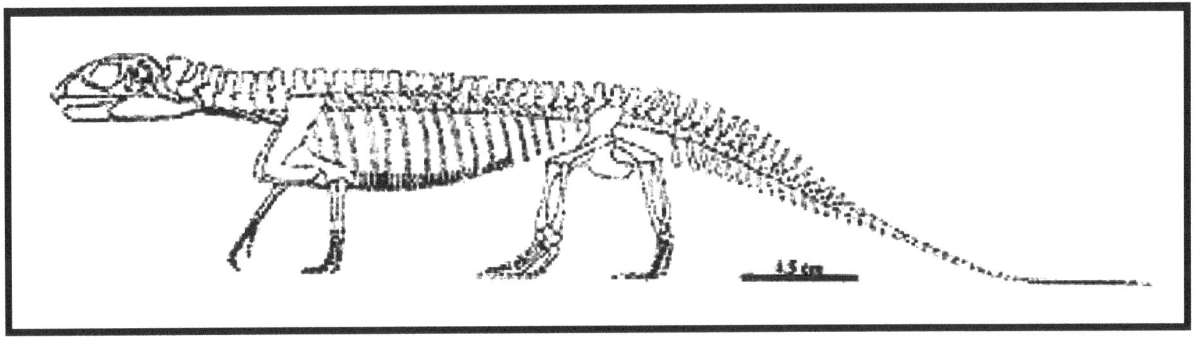

Paradapeton (order ryachosaurs)

On dry land the *diapsids* originally a*mbulated* in the same *sprawling* manner as *lizards*, with *extended* limbs *spreading out* from the *body*, moving them *horizontally* in *wide arches*, with their body *waving* from *side* to *side*.

But when they *swam,* however, they would hold their limbs *close* to their body so as to *minimize* the water *resistance*, and when *leaving* the water to *occupy* the banks, they eventually evolved *thighs* and *knees* in *close* proximity to the body *flanks*. They *walked* by *swinging* the *lower* parts of their *hind* legs *vertically*, with their feet *directly under* the body. They additionally evolved a *narrow-track gait*, which enabled them to *walk* on their *hind* legs alone, balancing the *forepart* of their body by their heavy muscular tail.

This gait proved *advantageous* as it as it required much *less* muscular effort than the *sprawling* style of walking of their *diapsid forebears*, resulting in a *key* change in the evolution of the *original* breast-pelvic skeleton. Since the *forelegs* were relieved from the burden of the *body-weight,* the *breast-skeleton*, apt for the strong muscular four- legged *sprawling* reptile, was *reduced* in the dinosaur design model, except in *birds* where it remains a *necessary* support for its on-going flight.

The *pelvis* consisted of the *iliac* bone, which was *attached* to the *spine* by a few vertebrae, the *pubic* bone, connected to the *iliac* bone, faced *forwards*, while the i*scium* bone faced *backwards*.

The bipedal, narrow-tracked gait, in which the pelvis *alone* should carry the *entire* w*eight* of the body, needed however, *more* vertebrae to *attach* the *spine* to the *iliac* bone, and consequently, was *elongated. Bipedal gait* also caused the muscles (which in the q*uadrupedal diapsids* had pulled sideways-extended thighs *downward)* to pull forward- extended thighs *downward*, as the *hind* legs carried the *entire* body-weight. As these muscles were *attached* to the *pubic* bones, they were elongated and bent *downward* in order to *reinforce* these muscles. This *pelvis* structure with the *long* ilium and the *long*, d*ownward* faced *pubis* is also characteristic of the bipedal *archosaurs* and *saurischians.*

As a consequence of the bipedal, narrow-tracked gait, the *archosaurs* could develop l*ong* hind legs. Later on, some of them also lifted their heels *off* the ground to become i*gitigrades*, which *changed* the ankle-joint into a simple hinge.

In the *Triassic* the climate became *more* humid, and the plants began to *spread*. The bipedal *archosaurs* followed as they *diversified* into *many* species, some as plant eating h*erbivores*, and some as meat-eating *carnivores*. Many of these *archosaurs*, however, eventually gave way to the *bipedal* walking dinosaurs.

The original *archosaurs* – the *phytosaurs* and the *aetosaurs*, were far smaller than the later dinosaurs. The modern crocodiles are more than likely to have originated from such q*uadrupedal archosaurs*, and have reassumed the same swimming style.

The ornitischian

As the earth was growing with plants, some of the bipedal *archosaurs* supplemented their carnivore *diet* with *plants* with *some* eventually becoming full-time *plant eaters*. Foliage trees had yet to *exist*, but a sort of tree which were abundant, were *ferns*, at **2-3** meters high, with *palm-like cycads*, and in order to *reach* the *sappiest* shoots of these trees some of the bipedal *archosau s* attained the habit of *raising* the *forepart* of the body to a s*teeple* angle, turning the thigh as long to the *rear* as possible, almost *perpendicular* to the spine. As an adaptation to its habit, its iliac-bone, which attached to the pelvis to the spine, was further elongated, and the downward-directed pubic bone was bent in behind and came to lie parallel to the *ischium.* This reinforced the muscles, which pull the thigh in behind and maintain the steeple position.

To further *adapt* as *plant eaters* they evolved front teeth in the upper-jaw and a tooth- less “pre-jaw” at the lower jaw, which made them better able to bite plants off. Their back teeth eventually changed to better chew the plants, while at the same time, also improving their digestion system. Thus the *ornitischian* group of dinosaurs arose, and as an improvement in their ability to walk, the pubic bone evolved a forward branch, almost parallel to the ilium.

The bipedal ornitichians

The original group of bipedal *ornitichians* supported on the entire side of their **five-toed** foot, but they gradually evolved *digitigrady*, the number of toes reduced to **three**.

The *ornithopods* group evolved into many species. Many new kinds of plants arose in the *Cretaceous*, attracting many *ornithopods* to adapt and feed on these plants. One of their group, the *hadrosaurs*, were needless to say, out of the ordinary, as they had a very *broad* mouths without any *front teeth*, with most of them developing *spectacular* crests on top of their heads. The purpose of these ornaments defies any explanation, but their close connection to the nasal passage points to a site for their *offactory* sense.

The Quadrupedal Oraitischians

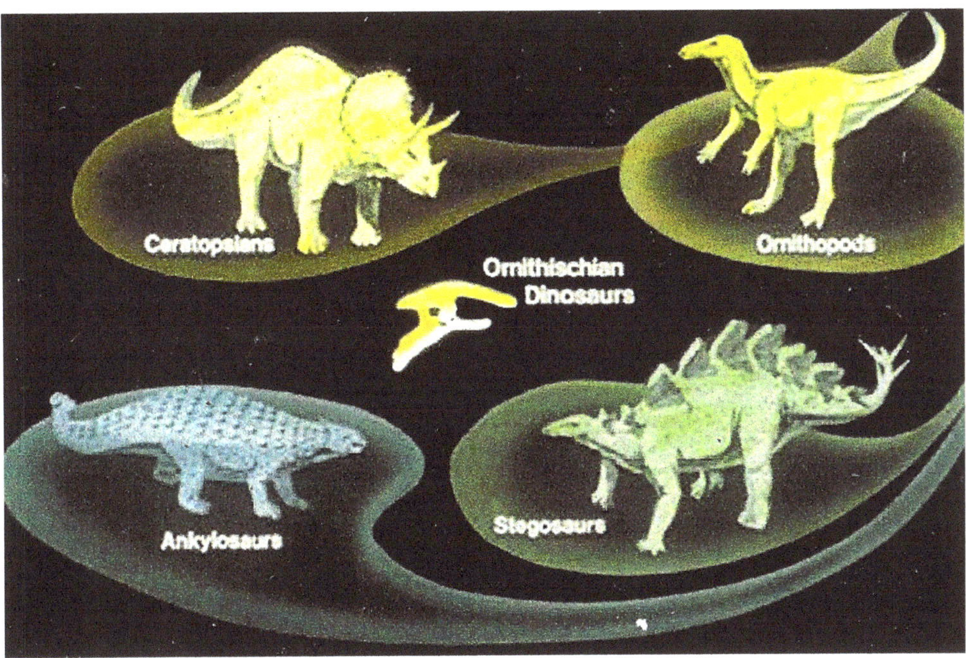

The stegosaurs and the ankyiosaurs

Some of the *ornitischians* specialized in eating lower parts of the plants and gave up the habit of raising the body. They, instead, supported the forepart of their body on the forelimbs and retained a quadrupedal way of walking. They did not evolve *digitigrady* and maintained the five toes on each foot. Those were the *stegosaurs* and the a*wkylosaurs*, which were provided with bony knots on the back. The *stegosaurs* had bony plates, which covered the neck and back, with bony plates on the *awkylosaurs*, covered the vertical bony back, like a shield. Without doubt, it stands to reason that these armaments had evolved to protect these low plant creatures against their taller predators.

The ceratopians

The four-footed *ceratopians* originated from bipedal *ornitischians*, which had acquired digitigrady, but gave up the bipedality before the number of **five toes** on each foot were r*educed*. They evolved a big bone collar, which covered the neck, and they are generally characterized by tall bony horns projecting from the front of the head in a rhinoceros-type fashion. This was doubtlessly a protection they had acquired after they had reassumed the q*uadrupedal* manner of walking.

The Saurischians

The quadrupedal sauropods

Instead of rising at their hind legs to reach high parts of the plants as the *ornitichians* did, another group of bipedal plant-eating *archosaurs* evolved a long neck. Thus they maintained the original structure of the pelvis with the pubic bone pointing downward.

The *quadrupedal sauropods* evolved a *peculiar* manner of *digesting*: instead of c*hewing* their food, they *swallowed* stones, which *grinded* the *vegetable parts* in their s*tomach*. They evolved a long tail, and as four-legged plant-eaters with a narrow gait, they could and did assume *gigantic* sizes. This same manner of *digestion* can also be found in *birds*.

The Bipedal Theropods

The carnosaurs

As the plants were abundant and *many* herbivores *arose* (including the plant-eating a*rchosaurs* as well as the plant-eating *rhynchosaurs*), there was basis for carnivores to establish themselves, with some of the bipedal *archosaurs* taking advantage of this opportunity, one of whom was the short-necked *Ornithosuchus*, of the Triassic.

They hardly needed to *chase* for the *kill*, but instead *lurked* around like bandits, with their *main course* likely *dead*, their prey (herbivores *ornitischians* and *sauropods*) usually easy prey – weakened or injured.

As these *tall* victims had a *thick* and *tough* hide, it was necessary for the carnivores to cut through the hides with their long razor-sharp teeth. Equipped with large jaws, the bipedal carnivore *archosaur caraosaurs* developed large heads and attained great heights.

The bipedal locomotion of the *carnosaurs* was *predicated* through their hind legs, which evolved *digitigrady* whose number of toes were *reduced* to *three* on each foot. The forelimbs, which were *not* used for walking were also *reduced* and as a result, became tiny.

Tyrannosaurus rex

The notorious **Tyrannosaurus rex**, was *without* question, the best known dinosaur, a *notorious villain*, who came to *symbolize* the *supreme* dinosaur *predator*, was as big as an *elephant*, and had *forelimbs* the size *only* of a man's *arm*. But the small forelimbs, which in *Tyrannosaurus rex* had *only* two fingers with sharp claws, with strong musculature that were not *totally* reduced, most likely because they were necessary for a big heavy male *mounting* a female for *mating*.

The Coelarosaurs

At an early age stage some long-necked *archosaurs,'* ancestors of the long-necked s*auropods*, added small *living* prey to their basically *vegetable* diet, and their long legs and long necks made them *fit* for *fast* moving and *rapid* snatching. Their sharp-pointed, t*hecodont* teeth, evolved for *fish* eating, which were also advantageous for them as ground-dwelling predators, and evolved *digitigrady* and long metatarsals. These bipedal predators evolved *numerous* species, and though very *agile*, they never became particularly tall.

At an early stage, before the original breast-skeleton was yet *reduced* – some very small, long-necked bipedal *archosaurs* got in the habit of climbing up tree trunks to escape their predators, and before long, they converted the *tree crowns* into their permanent habitat – these were called the **pro-avians** – ancestors of the **birds.**

The tree-climbing pro-avians

The trees, which the *pro-avians* used to climb, were not *branched* trees as present-day trees are, but palm-like *cycads* the flat crowns of which was suited for the *small* bipedal a*rchosaurs* to lay their eggs. Here they were safe from *egg-robbers*, and their hatchings, which were endangered by predators when living on the ground, were now much more secure.

The *pro-avians* performed tree-climbing, *step-by-step*, walking, hooking, and clinging by their *claws* on the tree-trunk with their outstretched forearms.

The *use* of the *forelimbs* demanded *strong* effort by the musculature, and therefore the p*ictorial* skeleton was *not* reduced as in the other bipedal *archosaurs*, but was maintained and developed. And as an adaptation to the tree-climbing behavior, their forearms were elongated and thus transformed into the *three-fingered hands*, with sharp claws and wrists to flex to the rear only in the lateral plane. This forearm structure, which is found in *birds* as the skeletal *framework* of the *wings*, was also found in the later *coelurosaurs.*

As the *ultimate* adaptation to the *tree-mounting* behavior the *pro-avians* evolved the f*lying* capacity to eventually become the *birds*.

Life in the tree crowns

The laying of eggs in the crowns of the *cycads* gave the small *pro-avians* and their hatchings a great advantage, but on the other hand, also a disadvantage insomuch as they were more *exposed* to the *heat* of the sun, with *little* or *no* shade available, thus becoming very *vulnerable* to *overheating*. As a means to *stay* as long as possible in their new tree-top safe heaven, their small *knobbed scales* evolved into a *plumage,* which yielded some protection against the intense rays of the sun; it served also as a means to r*educe* the body-temperature, together with air sacs which connected as a *coolant* to their lungs so as to *ease* their breathing.

The optimal body temperature for the *pro-avians* was *capped* at **42** degrees Centergrade, which proved to be *disadvantageous* to them when they sought food in cooler, shady places on the ground. In order to maintain this high body-temperature, they had to *increase* their *metabolism* rate in order to produce *more* body heat and so acquire the ability to maintain the necessary and consistent body-temperature needed to *evolve* into *warm-blooded* animals.

The bird-like coelurosaurs

The agile, warm-blooded *pro-avians* split into *two* main species, with many of these eventually *abandoning* the *tree-climbing* life-style and became *grounded* again as bipedal creatures. The breast-musculature, which was maintained for *tree climbing*, was then reduced – the more *reduced*, the more *time* elapsed in their giving up their tree-climbing, along with their tree-crown habitat.

These *defectors* from the bird lineage are the same *cueturosaurs*, although they're still very *bird-like* in appearance, and had somewhat reduced their breast-skeleton – but some, however, namely *Oviraptor* and *Felociraptor* had nevertheless *retained* a part of their clavicle.

The *long* three-fingered hands together with the *long* sharp claws were retained (although *reduced* somewhat with the passing of time) for other purposes despite the peculiar construction of their wrists.

As these offspring from the *avian* lineage no longer *needed* to *climb,* they grew taller, as they *split* into many *other* species of *various* sizes, but maintained, however, the *same* shape of the *theropod* dinosaurs. And although they were *lighter* in build, *together* with l*ong* necks, they were, nevertheless, more *so* than *not* bird-like in appearance. The d*escendents* of those who gave up *tree-climbing* at the *earlier* stage were *less* bird-like, however, than *those* who did so later.

The average sized *Coelophysis, Saltopus* and *Procompsognatus*, from the end of the Triassic, were *descendents* from *earlier* offspring lineage of the birds. The *later* they shed their tree-climbing ways, the *more* bird-like they became. The *small*, feathered *Archaeopteryx* from the Jurassic, with its long tail and toothed mouth, was without doubt a *near* pro-avian offspring, which, however, had *not* yet, however, reduced its *fused* clavicles and breastbone, whereas the small *Compsognates*, of Jurassic, were an *earlier* and more distant pro-avian offspring.

The *obvious* bird-like appearing *coelurosaurs,* from the *Late Jurassic* and *Early Cretaceous*, the toothless *ornithomimosaurs*, such as *Struhtomimus*, and *Ornithoomimus* were *descendents* from more *distant* pro-avian offspring, which had adopted a running, p*lant-eating* lifestyle.

Dinosaur relations

The dinosaurs, as we have learned, originated from a *group* of *reptiles,* in the *Early Triassic.* The *archosaus* (ancestor of dinosaurs), had *elongated jaws* with *thecodont* dentition and antorbital opening, a narrow-tracked gait, along with a *bipedal* style of walking. They eventually evolved into a *myriad* and *diverse* groups of both *carnivores*, h*erbivores*, and *omnivore* (*in-betweens)*, coming in all *shapes* and *sizes*, living in relative c*lose* proximity to one another.

As an illustration of diversity between *dinosaurs*, the *toothless* bird-like o*rnithomimosaurs,* for example, stands in *stark* contrast to *Tyrannosaurus rex*, generously *equipped* with a *large* mouthful of ready-to-eat large *razor-sharp teeth.*

The herbivores came in both *short-necked* and *long-necked* varieties. The short- necked ones had a general *change* in *pelvis*, and a *dentition* for chewing food. They became the *ornitichian* dinosaurs, which split up into *ornithopods*, but remaining *bipedal*, whereas the *stegosaurs* and *ankyluosaurs* subsequently assumed *quadrupedality*, with the c*eratopsians* doing so later.

The *long-necked* herbivores were the *ancestors* of the tall, bipedal *prosauropods*, *some* of which were the ancestors of the huge *quadrupedal sauropods*. Some early ancestors of the *long-necked* herbivores were also the root of the bird-linage, the long- necked, *lightly* built *pro-avians*. From the lineage of *pro-avians* the early *coelurosaurs* branched off, and eventually the *crnithomisaurs* and *deinoychosaurs* **defected** from the bird-lineage.

From the bipedal *archosaurs* the tall, short-necked meat-eating *carnivores* emanated separately. An efficient *eating-machine*, with a myriad of *tools* at hand to do the job, which included among features: *second toes*, along with *sharp razor sharp teeth* to *catch*, h*old* the prey and with his *claws*, which able to *cut* through the *toughest* hides and then carve it up into *separate delicious morsels.*

The dinosaurs' extinction

As for the *extinction* of the dinosaurs, the only *solid* fact is that *no* trace of dinosaur has been found *after* the Cretaceous. This no doubt *calls* for so some creative *inquiry* of their s*udden* demise, along with a *myriad* of theories, many of which, however, have been already proposed, *mainly* though – only for *dramatic effect*.

It certainly goes without saying, that the dinosaurs were a *very* successful species, the d*ominant* ones from the *Triassic* to the *Cretaceous periods*, with *little* to *no* serious c*ompetition* from *any* of their *contemporaries*. The mammals, *as we know them*, were o*nly* little *insignificant* furry *nocturnal tree dwellers* during that time.

At the *beginning* of the *Triassic, various* other groups of *reptiles*, which had survived the *great* **Permian Extinction**, *reestablished* their dominance in the ecological system of the *Mesozoic*. These were the **diapsid reptiles**, which evolved into the *quadrupedal*, s*prawling* lizards, and the herbivore *fhynchosaurs.* The *former* (diapsid reptiles) remained *small*, with the latter (fhynchosaurs), *slow, clumsy* animals, died out during the *Mesozoic*. The **birds** *arose* during the *reign* of the *dinosaurs*, and so also did the *flying* **r***eptiles*, the **ptertosaurs**, amphibian *archosaurs, phytosaurs*, and *crocodiles,* and the various species of *sea-dwelling reptiles*, the *ichthyosaurs* and the *plesiosaurs* – but none of these, however, proved to present any serious *threat* to the dinosaur domination. From the *Middle Triassic* the stage was *indelibly set* for the reigning dinosaurs, as they m*andated* an *eco-system* of their very *own*.

This system was, more or less, *based* on the *tall* herbivore plant-eating *sauropods*, the Various *ornitischians*, and eventually on the *bird-like*, *ornithomimians*. These plant- eaters drifted all over the *limitless* landscape, as *hosts*, if you will, for *lack* of a *better* w*ord*, consuming an enormous, *eye-boggling* amount of plants, which *translated* into a *bonanza* of large amounts of *meat* to the *benefit* of their *uninvited guests*, the never too far behind escorting *carnivore* meat-eaters.

These *plant-eating* herbivores were objects of *constant surveillance* and stalking by the *meat-eating* carnivores, among which, were the bird-like *coelurosaurs*, all in great anticipation of the *tasty* feasts of the huge feast awaiting them, when one of their generous large, *meat-loaded* hosts lay down for good, and made no difference whatever the reason for the demise – natural, injury, etc., or attacking young or invalid ones who could be easily overtaken – but even so, there is no reason to believe that the meat-eaters were primarily aggressive, attacking predators. More likely they were content to be scavengers, which made the acquiring food much easier than pursuing their, always an option, equipped as they were with their large, efficient *eating tools*.

Among the things dinosaurs owed their superiority to, was their *narrow gait*, which they had *achieved* from their *swimming* and *bipedal* ancestral *archosaurs*. The gait allowed them to evolve *long legs* and *move* in a manner, which required much less energy than the *sprawling* gait, followed a much improved gait of the *pre-mammal Synapsid/Therapsids.*

Synapsid/Thrapsids.
Because of their *superiority* and *dominance* over their *contemporaries*, there was generally *no need* for any *major* evolutionary changes in their physical design, or breeding habits, during their *long reign*. Conversely, this was *not* the case, however, on the part their *rapid* up-and-coming *mammalian* contemporaries.

Although *some* dinosaur species had room for improvement concerning their d*igestion* and *metabolism* rate, which could have been *higher*, with some degree of e*ndothermy*, along with some *improvement* with *watching* over their offspring more carefully, *dinosours* of the Late Jurassic and Early Cretaceous, *defectors* from the bird lineage, were *warm-blooded* with a *high metabolism* rate and of a somewhat an a*ggressive* bent. For the most part, however, the dinosaurs *remained* basically reptiles with a *low* metabolism rate, and rather *docile*. Because their poor *eyesight, their* ability to recognize objects in their immediate surroundings was nothing to *brag* about, resulting in their sense of *disorientation*. Their hearing, although not up to par, proved adequate.

The ornamented crests of the late *ornthopods*, however, was said to have given them some ability to orientate through their sense of smelling. It is generally *conceded* that the dinosaurs were neither *above* nor *below* the intelligence of their contemporaries, except, of course, **mammals.**

The *Mesozoic* witnessed the explosion of the dinosaur *numbers*, each one of which species adapted its own *niche* in the environment, and by the end of the *Cretaceous* the entire dinosaurian world was one great **dinorama**.

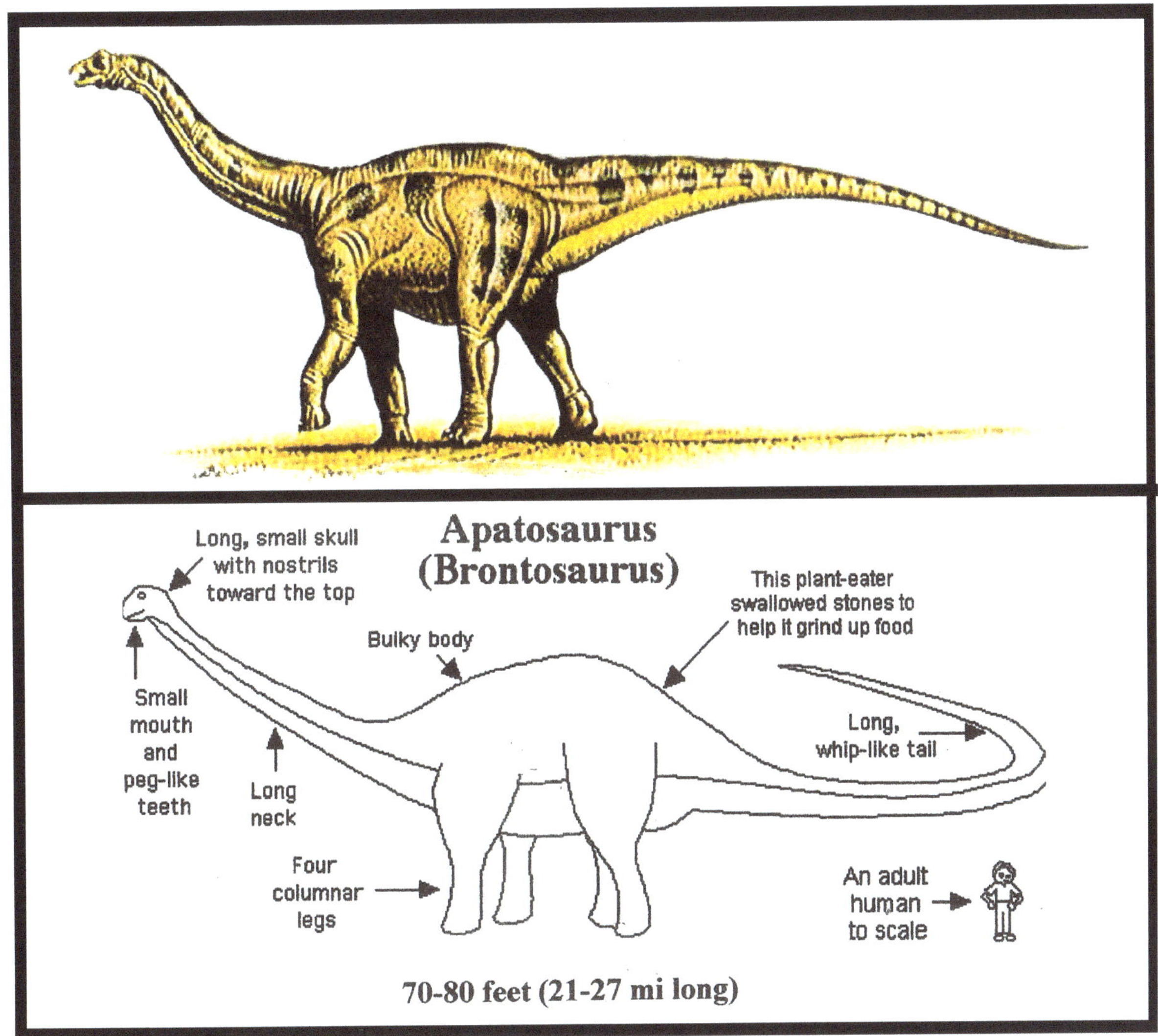

Their large size, reaching dizzying heights in *some* species, along with their *long* legs, in effect, confined them to the open, *flat-lands*, making them unsuitable to *rocky, swampy* and *dense* forest landscapes, not unlike modern large, long-legged mammals.

During a greater part of the *Mesozoic*, however, this posed no disadvantage to the mammals.

Great herds of *plant-eating* dinosaurs *persisted* on their all-consuming wanderings over the *open* landscape on their *quest* for plants, eating just about *all* the greenery in their paths, and leaving a *devastated* area in their wake.

When the *day* of these particular dinosaurs had eventually *passed*, the *ground* was left with only bare, dying, and crushed trunks, along with heaps of dung. Then the *cycle* would start *all over again*, again as *new* plants would eventually grow again, until the next *herd* of plant-eaters *emerged* again, *stalked* as usual, by their meat-eating *scavenger* carnivores.

The *foliage* plants and *foliage* trees evolved in the *Cretaceous*, but the plants, however, did not suffer any *ill-effects* in having their green leaves *bitten off*, since they simply grew *new* foliage, thus resulting in subsequent creation of *new* forests, and causing the further *reduction* of space for the dinosaurs.

Although the *balance* of the *ecosystem* was *upset* – so far as the dinosaurs were concerned – along with some *decreased* plant-eaters – it did not, however, have a l*asting* impact on them. But, nevertheless, by the end of the *Cretaceous*, the entire ecosystem *collapsed*. Because of their *specialization* as plant-eaters, dwelling in a limitless landscape area, and now contained in a much smaller environment, proved to be something not only too *physically* restricting for them, but also *alien* to their *habitual senses* (multiple millions of years in the making) to adjust to in the *sudden* and *traumatic new* situation that *pounced* on them. Finally, *relatively* shortly thereafter, they disappeared for good, leaving the earth *free again* to be covered with a carpet of u*nmolested* pristine **greenery**.

This *extinction* also resulted some beneficial spin-offs, in that the amount of *carbon dioxide* in the *atmosphere* was *less* than **1%**, but the *plants* lived on it, *attaching* it as c*ellulose*, thereby *removing* it from the atmosphere. But as the *herds* of dinosaurs had eaten the plant-life with *hungry abandon* and without *discretion*, the attached **CO 2** was constantly being *released* into the atmosphere. Consequently, as the forests began to s*pread* following the *absence* of the dinosaurs, the *great amounts* of *attached* **CO 2** were no longer beng *released* resulting in a *reduction* in the amount of **CO 2** in the atmosphere.

In the **sea**, the *algae* also *depended* on the **CO 2**. During the *Cretaceous* the algae that a*ttached* it and then *disposed* of it as *chalk*, consummated *enormous* amounts of **CO 2**.

But as the *increasing* forests *attached* still *larger* amounts of **CO 2**, and the atmospheric amount of **CO 2** *fell*, the algae of the sea *decreased* which, together with the *dinosaurian effect*, also *upset* the ecological *balance* of the sea. The *algae* was *food* for other sea organisms – resulting in a *decrease* in fish population, along with a *myriad* of other sea creatures. This also helps to *explain* why our *ancestral* **lobe-finned fishes**, *along with sea-reptiles, ichthyosaurs plesiosaurs, mesosurs*, and *many others* eventually d*isappeared*.

All of these factors, along with others (one of which is attributed mainly to a comet hit on earth, which continues to gain *popular* support, along with some *scientific* support), had contributed to the dinosaurs' **end.** It also goes without saying that the *greatest contribution*, so far as **we** are *concerned*, is that it *freed* **our** ancestors, the small, *warm- blooded mammals*, the *opportunity* to *evolve*, leading, of course, eventually to **us!**

INTRODUCING SAURICHIAN DINOSAURS

Dinosaurs are divided into groups based on hip structure, the **Order Ornthischia** and **Order Saurischia** The *Saurischian* dinosaurs (*Order Saurischia*) were the ancestors of birds. They had a hip structure similar to that of lizards – the pubis bone pointed downwards and forwards. *Saurischian* skulls also have large, pronounced openings (*antorbital fenestrae*) between the eye socket and the *nares* (*nostrils*), much more so than the *ornithishians*. The *Saurischians* are divided into the *sauropodmorpha* – four-legged herbivores, and the *theropods* – two-legged carnivores. The oldest known dinosaurs (*Eoraptor* and *Herrerasaurus*) are *Saurischians*, and date from the mid-to-late *Triassic* period, about **230 million years ago**.

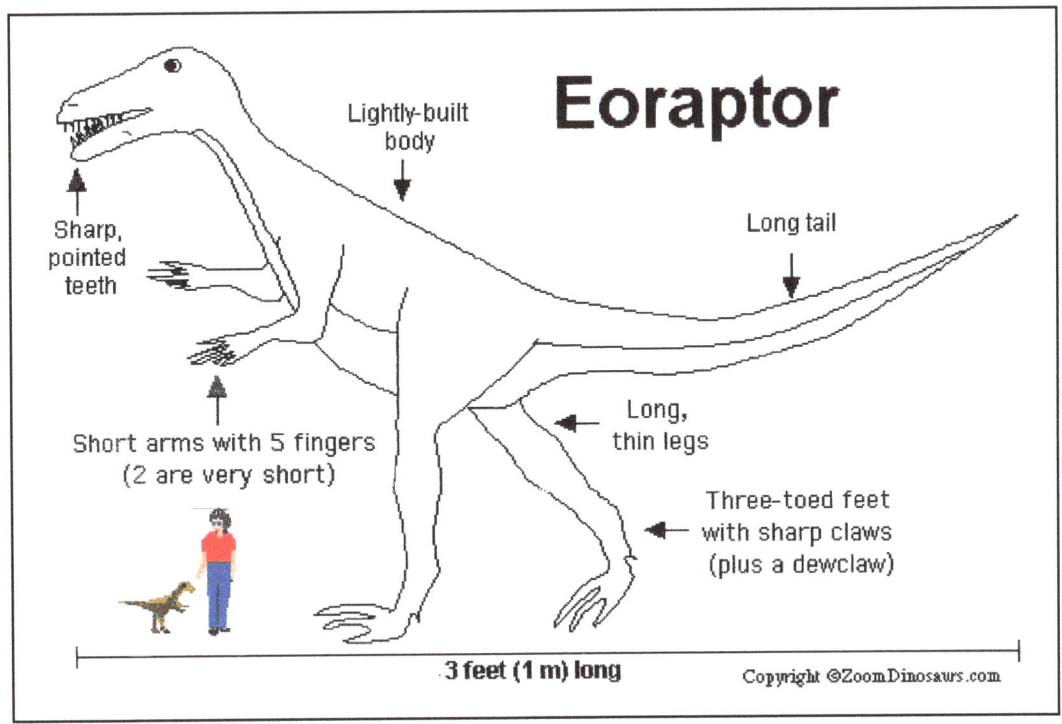

Major divisions of the Saurischian dinosaurs

Sauropodomorpha

* Plant eaters

* Tiny head and brain

* Big gut for digestion of lots of plant material

* Long neck for grazing

* Four columnar legs

* Tail for counterbalancing the long neck

* Small heads with spatulate or pencil-shaped teeth

* Usually walked on four legs

* Largest land animal ever

Theropods – Led to birds

* Carnivorous diet

* Fast and agile

* Sharp, slicing teeth (or beak), and well-developed jaw muscles

* Clawed hands: usually with three main digits – exceptions include the

Tyrannosaurids, who *lost* the *third* digit

* Bipedal walk

* Strong legs with bird-like, clawed feet

* Large eyes, indicating good eyesight.

* Short arms for grasping prey

* Tail for counterbalancing the neck and head

Long rear legs for speed

The following is a sampling of Surischian dinosaurs

Prosauropods – 4 legged herbivores

Massopondylus	Plateosaurus	Riojasasaurus
Amargasaurus	Apatosaurus	Brachiosaurus
Camarasaurus	Diplodocus	Supersaurus
	Ultrasauros	

Theropoda—"Beast-footed," 2—Legged Carnivores, Evolved into Birds

Albertosaurus	Allosaurus	Baryonyx
Carcharodontosaurus	Caudipteryx	Coelophysis
Compsognathus	Deinonychus (antirrhopus)	Dilophosaurus
Foraptor	Gallimimus	Giganotosaurus
Megalosaurus	Ornitholestes	Ornithomimus
Oviraptor	Protarchaeopteryx	Saltopus
Scipionyx	Senosauropteryx	Spinosaurus
Troodon	Tyrannosaurus rex	Unenlagia

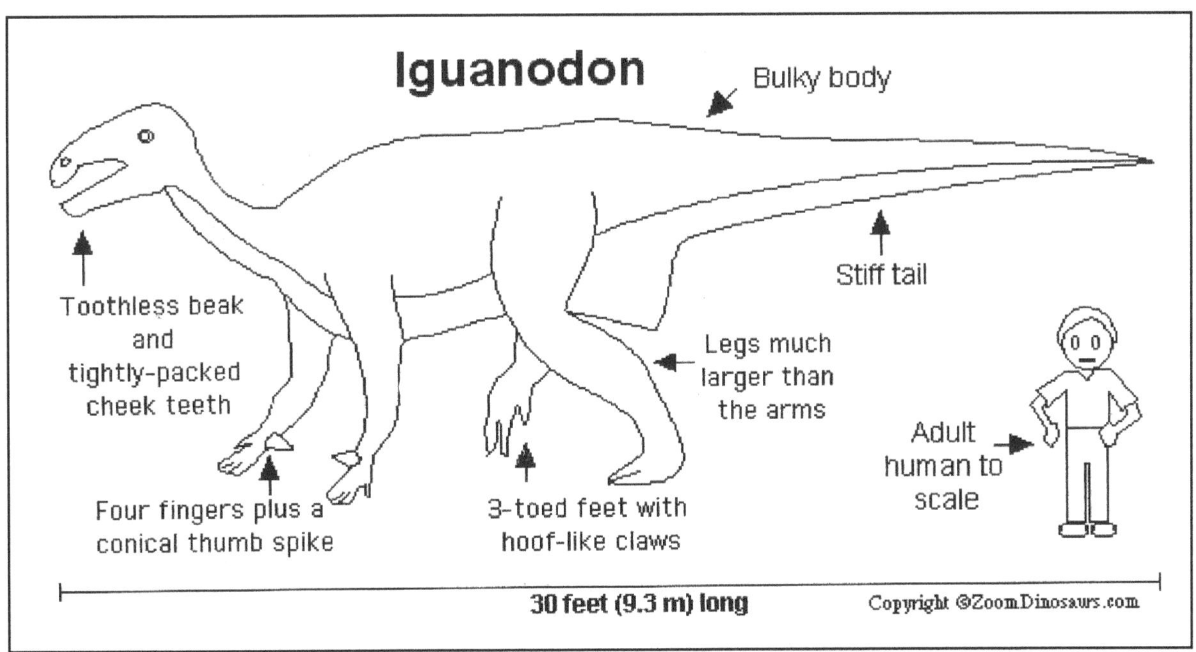

Iguanodons *lived in the early Cretaceous Period, about* **135-125 million years ago**, *towards the end of the* **Mesozoic Era**.

INTRODUCING THE PTEROSAURIA

The Flying Reptiles

Ranging from the size of a *sparrow* to the size of an *airplane,* the **pterosaurs** (*Greek* for "**wing lizards**") *ruled* the skies in the *Jurassic* and *Cretaceous*, and included the *largest vertebrate* ever known to *fly*: the late *Cretaceous Quetzalcoatlus*. ***Note***: The *appearance* of **flight** in *pterosaur* is *separate* from the *evolution* of **flight** in *birds – pterosaurs*' are n*ot* even *closely related* to either *birds* or *bats*, and thus provide a classic example of co*nvenient* evolution jargon.

It was once thought that *pterosaurs*' were not well adapted for *regular* flight and relied largely on *gliding* and on the *wind* to stay in the air. Based on analyses of p*terosaurs*' skeletal features (including the work done by Berkeley's own *Keven Padian*), it is now thought, however, that all but the largest *pterosaurs* were able to *sustain powered* flight.

The largest *pterosaur (Quetzalcoatlus*, wonderfully named for the Aztec winged s*erpent* God) had a wing span from **eleven** to **twelve** meters long (roughly **forty feet**).

The wing's main support was an amazingly elongated *fourth* digit in the hand. Fibers in the wing *membrane* added structural *support* and *stiffness*. At least some *pterosaurs* may have had some sort of *hair-like* body covering, which could very well have meant that they were *endothermic*.

Ptercosaurs had a *diverse* range of head types, as so you can see from the pictures shown. Their ability to fly allowed them probably to *evolve* into a *myriad* of niches, no doubt taking of advantage of the *opportunities* presented by the different *food* sources, which would explain the range of the *skull* morphology seen.

Pterosaurs consisted of *two main types* (they were, however, *basically* a *single* m*onophyletic* group) – the "rhamphorhynchoids," more properly termed, the *basal Pterosaurs, with long tails*, with their descendents, the "pterodactyloids," with shorter tails. Since the later *Pterosaurs ("pterodactnchod*") are the descendents of the b*asal pterosauria*, the term "*rhamphorhynchod*" is invalid. "*rhamphorhynchoid*" is a p*araphyletic* term, which *phylogenetic* researchers shy away from using. The *basel Pterosauria* (including Rhamphorhynchus) first appeared in the Late *Triassic*, going e*xtinct* at the end of the *Jurassic*.

Pteranodon

The more derived p*terosaurs* (including p*teranodon,* as shown) that were the d*escedents* of this group appeared first in *Late Jurassic* rocks, and the last of them died out at the *end* of the *Cretaceous*. Shown is a picture of a mounted skeleton of *Pteranodon ingens* on display at the UCMP.

What Was Pternadon Like?

The genus *Pteranodon* includes several species of large *pterosaurs* from the *Cretaceous Period* discovered in North America. As you can tell from this photo, it had a *large* crested head, a *huge* wingspan (some **20-25 feet**: the UCMP specimen is about **22 feet**), and a comparatively *small* body. This *appears* to be *deceiving* at *first glance*; it looks as if the head and wing *bones* are *too bulky*, and the hind-limbs appear *small* and *weak*. Not so the *case*: the *bones* of *Pteranodon* were actually *hollow* (about *1 millimeter thick*!), and quite *light*. The whole creature *weighed* around *25 pounds*, just slightly *heavier* than the *largest Modern birds*. The *hind-limbs* were actually *perfectly sized* for the body.

Pterandon would have been capable of *bipedal* terrestrial movement – but was Not, relatively speaking, a *rapid runner*, unlike *some* its *ancestors*. The wing bones appeared t*hick* since *large diameter bones* are *vital* for *resisting* the *bending* stresses involved in flight, as opposed to *large bone thickness*, that are important for *resisting* compressing forces, such as those *imposed* by the *weight* of a *large* body – so given all of this, the wings of *Pteranodon* would seem to be *more* than *adequate* for flight.

It can probably be accurately assumed that *Pteranodon* was a souring creature – it used *rising* warm *air lifts* to maintain altitude, a common strategy employed among *large* modern birds – *albatrosses* and *vultures*. Its scoop-like beak was utilized for snapping up f*ish* as it *soured* over the *oceans* that it also *nestled* by. A good *analogous* for *Pteranodon* would be the *modern* fish-eater – the *pelican*.

INTRODUCTING THE ICHTHYOSAURIA

Ichthyosaurs

While dinosaurs ruled the land, **ichthyosaurs**, classified variously in **Ichthyosaurs** or in the **Ichthyoptergia**, shared the sea with the other great groups of *large* marine *reptiles*, the *plesiosaurs* and *mosasaurs*. "*Ichthyosurs*" means "fish lizard," while "*Ichthyoptergia*" means "fish paddle." Both are apt. The earliest *ichthyosaurs* had *long, flexible bodies* and probably *swam* by *undulating*, like *living eels*. More advanced i*chthyosaurs* – like the one shown here, on display at the *Senckenberg Museum* in *Frankfurt, Germany* – had compact, very *fishlike* bodies with crescent-shaped tails. The s*hape* of these *ichthyosaur* is somewhat like that of living *tunas* and *mackerels*, which are the fastest fish in the ocean – and like them, the latter *ichthyosaurs* were built for speed. **Note**: the *paddles* with which *ichthyosaurs* swam, had the same basic *plan* as your h*and* and *arm,* but the arm bones, however, were very *short*, while the *fingers* had lengthened by developing many more *bones* than the *three* that make up each of *your* fingers.

Rare fossils have been found that show *ichthyosaurs* actually giving *birth* to its live, and *well-developed* young – *ichthyosaurs* never had to *leave* the water to lay eggs. In fact, *judging* from their streamlined, *fishlike* bodies, it seems almost certain that i*chthyosaurs* could not *leave* the water. Yet still they **breathed air***,* while at the same time, *lacking gills*, like *modern whales*.

Ichthyosaurs were **not** dinosaurs, but represent a *separate* group of *marine* vertebrates.

Because *ichthyosaurs* were so *specialized* and *modified* for life in the ocean, we don't really know *which* group of *vertebrates* were their *closest* relatives. They might have been an *offshoot* of the *diapsids* – the *great* vertebrate group that includes the *dinosaurs* and *birds*, the *pterosaurs*, the *lizards* and *snakes*, among many *other* vertebrates. On the other hand, some have suggested that the *ichthyosaurs* were *descended* from a distant relative of the *turtles.*

The first *ichthyosaurs* appeared in the *Triassic*. In the *Jurassic, ichthyosaurs* disappeared, however, in the *Cretaceous* – several million years before the last dinosaurs died out. Whatever caused the *extinction* of the *dinosaurs* did **not** cause the *ichthyosaurs* to also *die out*.

INTRODUCING THE ARCHAECOPTERYX

Urvogel – German name for ("original bird" or "first bird")

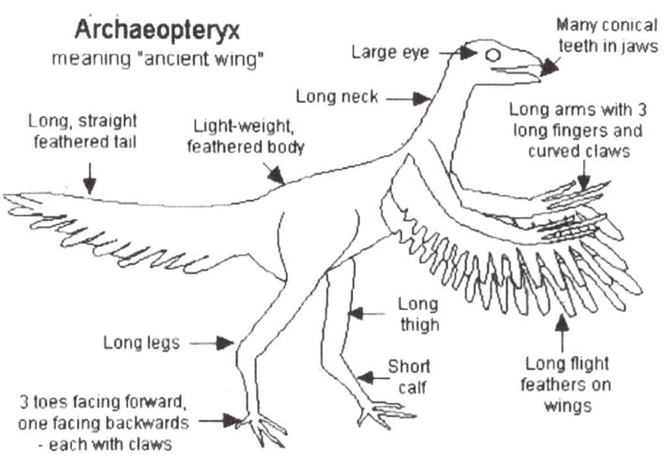

Archaeopteryx was the *earliest* and most *primitive* bird known. The name is from the Ancient Greek dpxaioc (*archios*) meaning '*ancient,*' Atepus (*pteryx*) meaning '*feather*' or '*wing*' (pronounced "*AR-kee-OP-ter-iks*"). *Achaeopteryxs* lived in the *Late Jurassic* around **155-150 million years ago**, in what is now *southern* Germany during the time when *Europe* was an *archipelago* of *islands* in a shallow warm *tropical* sea, much *closer* to the *equator* than now. Similar in *size* and *shape* to a *European Magpie, Archaeopteryx* could grow to about **0.5 meters (1.6 ft.)** in *length*. Despite its small size, *broad wing*, and ability to *fly*, *Archaeoptryx* had more in *common* with small *theropod* dinosaurs than it had with modern *birds*. It particularly *shared* the following features with the **deinonychosaurs** *(dromaeosaurs* and *troodontids*) – jaws with *sharp* teeth, *three* fingers with claws, a long bony tail, *hyperextensible* second toes ("killing claw"), feathers (which also suggest h*omeothermy*), and various *similar* skeletal features.

The features (shown here) make **Archaeopteryx** the *first* clear *candidate* for a t*ransitional* fossil between *dinosaurs* and *birds*, but also played the *pivotal* role not only in the *study* of the *origin* of birds but also in the *study* of dinosaurs.

The *first* complete specimen of *Archaeopteryx* was announced in **1862**, just two years after **Charles Darwin** published **On the Origin of Species**, which, of course, became the k*ey* source evidence in the *debate* on **evolution**. Over the years, nine more fossils of *Archaeopteryx* have surfaced. Despite *the variations on a theme* concerning these fossils, most experts regard all the *remains* that have been thus far discovered to be *those* of a single species – though, at this writing, however, the *debate* is still *on-going*.

Many of these eleven fossils include impressions of *feathers* – among the *oldest* (if not the *oldest*) direct *evidence* of feathers. Moreover, because these feathers are an advanced form (*typical* flight feathers), these fossils are plain *evidence* that *feathers* had been e*volving* for quite a while.

Archaeopteryx was the size of a medium-sized modern-day *bird*, with wings that were rounded at the ends with a long tail equal to its body length – measuring **500** *millimeters*.

Its feathers, although *less* documented than many of its other *features*, were *similar* to that of modern-day *birds*. Although appearing very much a bird, *Archaeopteryx* still retained a number of *theropod dinosaur* features. Unlike modern-day birds,

Archaeopteryx still had *small teeth* as well as a long bony *tail*, features *Archaeopteryx* shared with other *dinosaurs*.

Because it *displayed* features *common* to both *dinosaurs* and *birds*, *Archaeopteryx* has generally been considered the **in-between link**, and in all likelihood, the *first bird* to transcend from a *land-dweller* to a *flying creature* – a **bird**. In the **1970s**, *John Ostron*, following *T.H. Huxley's* lead in **1868**, argued that *birds* evolved from *theropod dinosaurs* and that *Archaeopteryx* was a *critical link* in the *evidence supporting* this argument, preserving features akin to both *theropod dinosaurs* and *birds*. *Ostron* found that *Archaeopteryx* was remarkably similar to the *theropod* family *Dromaeosauridae*.

The first fossil-find of *Archaeopteryx* was discovered in **1861**, only two years after *Charles Darwin* published *On Origin of the Species*. *Archaeopteryx* seemed to co*nfirm* Darwin's theories, and has since keyed-in the evidence on the *origin* of the birds' t*ransitional fossil debate*, further *confirming* and *bolstering* the *theory of evolution*. Subsequent research on dinosaurs from the *Gobi Desert* and *China* has unlocked more s*tunning* new *evidence* on the *Archaeopteryx – theropod dinosaur Link*.

First Dinosaur Found With Its Body Covering Intact Discovery of Remarkably Preserved Fossil Dinosaur from China on Display at American Museum of Natural History

New York . . . April 25, 2001 . . . A team of *Chinese* and *American* scientists announced today in *Nation* the *discovery* of a remarkably preserved **130-million-year old** fossil d*inosaur* covered from head to tail with *downy fluff* and *primitive feathers*. It is the *first* dinosaur found with its *entire body* covered **intact**, providing the *best evidence* yet that a*nimals* developed *feathers* for *warmth* before they could **fly**.

The dinosaur was *unearthed* last spring by farmers digging in the famous fossil beds of *northeastern* China's *Liaoning Province*. It is described in the science journal *Nature* by a team led by Ji Qiang, of the *Chinese Academy of Geological Sciences*, and *Mark Norell*, Chairman of the *Division of Paleontology* at the *American Museum of Natural History*.

The researchers have identified the fossil as **dromaeosaur**, a *small, fast-running dinosaur,* closely related to a group of dinosaurs known as *advanced theropods*, two- legged *predators* (in the same *group* as *Tyrannaosaurs rex)*, with *sharp teeth* and *bones* strikingly *similar* to those of modern-day *birds*.

"This fossil *radically modifies* our *vision* of these *extinct* animals," said *Dr. Norell*, whose discoveries in the *Gobi Desert* of Mongolia have led to *new ideas* about *theropods* and *bird origins* – It shows us that advanced *theropod* dinosaurs may have *looked* more like *birds* than giant *lizards*.

Entombed in two slabs of fine-grained rock, the dinosaur's skeleton *resembles* that of a *large duck* with a *long tail* and a *oversized head* (indicating that the animal was a *juvenile*). A *small fish* was embedded in the rock near the left foot. Its head and tail were covered with *downy fibers*. Other parts of its body sprouted *tufts* or *sprays* of *filaments* resembling primitive *feathers*, and the *backs* of its *arms* were *adorned* with branched *structures* like the *barbs* of a modern *bird feather*.

Since **1995**, when the *first* dinosaur with *primitive* feathers, *Sinosauropteryx*, was discovered in the *Yixian Formation* of the *Liaoning* fossil beds, several *new species* of dinosaurs with *featherlike* structures have been found there. But in most cases the fossils were *jumbled up* or *incomplete* – making it *unclear* how the *featherlike* structures related to the animal's *body*. Critics of the widely accepted theory that *modern birds* evolved from dinosaurs have *questioned* the *validity* of these "feathered" dinosaurs, claiming that the *featherlike* structures were *not* primitive *feathers* or that the specimens were mixed-up fossils of primitive *birds* and *dinosaurs*.

The detail on the *newly* discovered *dromaeosaurs* is so fine that it allows scientists to see how the primitive feathers were attached to the dinosaur's body. "This is the specimen we've been waiting for," said Ji Qiang. "It makes it *indisputable* that a body covering *similar* to *feathers* was present in *non-avian* dinosaurs.

Because *dromaeosaurs* are more *primitive* than *birds*, this fossil helps make the case that feathers developed **before** flight. In small, *flightless* dinosaurs like this one, feathers may have *evolved* as an *essential* piece of equipment for staying **warm**.

"Modern birds are *warm-blooded* and their feathers play an integral role in keeping them *warm,* so it is not unreasonable to state that *non—avian* dinosaurs developed *primitive feathers* at the **same time** that they developed *warm-bloodedness*," says *Norell*.

"It's not inconceivable that smaller dinosaurs like this one, and even the young of *larger* species like *Tyrannosaurs rex,* may have *needed* featherlike body coverings to maintain their *body temperature*.

Scientists have *yet* to determine if the *new species* represents a n*ew one*. But they do know that it *shares* some *anatomical* characteristics with the *two* other *dromaeosaurs* discovered in the s*ame* fossil beds – *Sinornithosaur,* as shown at left, a small dinosaur, first described in *1999*, and *Micoriaptor*, the *smallest* known *theropod* dinosaur *yet*, was found last year, in 2000. **A Treasure Trove of Fossils in China** Consisting of layers of volcanic and sedimentary rock, the *Yixian Formation* in China's *Liaoning Province* has yielded an *enormous* variety of *fossil fish, birds, insects, reptiles, shrimp, flowers, mammals* of all sorts, along with dinosaurs, dating back to the *Late Jurassic* and *Early Cretaceous* times, between **145 to 120 million years** hence. At that *time*,

Sinornithosaurus

the region was *dotted* with freshwater *lakes* and *volcanoes*. Volcanic *explosions rained* fine ash into the *lakes*, and animals that were *trapped*, fell into the *wate*, were immediately b*uried* into the *fine-grained* sediment at the *bottom*. Because they were buried so *quickly*, with so *little* oxygen *available* to initiate *decay*, the fossil animals found in the *Yixian Formation* had their *delicate* features miraculously *intact*, going from *feathers* to *fish* scales, to patterns on insect *wings*, etc.

"These fossils have dramatically *changed* the way we now *understand* what life was l*ike* during the *Late Jurassic* and *Early Cretaceous* times," said Ji Qiang.

How Are Dinosaurs Related to Birds?

In the last two decades, other *bird-like* dinosaurs and *dinosaur-like* birds, have been unearthed at fossil sites around the world, including those in *Madagascar, Mongolia,*

Patagoria, and *Spain*, etc. *Together* with the Chinese fossils, they *provide* the *strongest* evidence *yet*, that birds *evolved* from **therapod dinosaurs**. But, however, an *insignificant minority* of *dogmatic* scientists continue to *argue* to the *contrary* on the *dinosaur-bird link*, that birds evolved *independently* from *earlier*, yet undiscovered, *reptiles*, pushing them even *further back* in time.

The *link* between *dinosaurs* and *birds* first noted in the **mid-1820s** by naturalist *Thomas Henry Huxley*, who *observed* that birds were still much like *reptiles*, but with a noticeable *difference*, a **beak** instead of **teeth**, along with *three* reptilian **fingers** hidden i*nside* of their *wings*. In the **1970s**, *John Ostron* of *Yale University* launched a meticulous s*tudy* on *comparing* the *anatomical* features of dinosaurs and the *oldest* known bird,

Archaepteryx.

Today we know that *theropod* dinosaurs and *birds* share more than **100** anatomical f*eatures*, including a *wishbone, swiveling wrists*, and *three* forward-pointing *toes*. Among all advanced *theropods*, the swift-running *dromaeosaurs* are thought to be the most c*losely* related to *birds*.

INTRODUCING THE FIRST PRIMATES

Primates are remarkably *recent* animals. Most animal species *flourished* and became e*xtinct* long before the first *monkeys* and their *prosimian* ancestors *evolved*. While the *Earth* is about **4.54 billion years old** and the *first* life dates to at least **3.5 billion years ago**, the *first primates* did not appear until around **50-55 million years ago**. That was a*fter* the *dinosaurs* had become **extinct**. *Transitional* primate-like creatures were evolving by the end of the *Mesozoic Era* (ca, **65.5 million years ago**). At that time, the world was very different from today. The c*ontinents* were in *other locations* and they had somewhat *different* shapes. North America was still *connected* to *Europe,* **not** to *South America*. India was not yet *part* of *Asia* but heading towards it at a surprising rapid rate of nearly **8 inches (20 cm) per year**. *Australia* was close to *Antarctica*. Most land masses had *warm tropical* or s*ubtropical* climates.

The *flora* and *fauna* at the end of the *Mesozoic Era* (**245-66.4 million years ago**) would have seemed *alien* since most of the *plants* and *animals* that are familiar to us had not yet *evolved*. Large *reptiles* were beginning to be *replaced* by *mammals* as the d*ominant* large land animals. Among these mammals, were a few *archaic* egg-layers (monotremes) like the *ancestors* of the modern *platypus* and *echidna*. There were larger numbers of pouched *opossum-like* mammals (marsupials). The few *placental* mammals that existed at that time consisted mainly of the **insectivore** – ancestors of *primates*. The large grass-eating placental mammals, such as *cattle* and *wildebeest*, were absent as were the *vast* grasslands that would later develop. *Rodents* and seed-eating *birds* were also absent. The great *proliferation* of flowering *plants* had *not* taken place yet. Forests of broad-leafed trees were, however, developing, all over the Earth.

Primate-like mammals

The first *primate-like* mammals, or *proto-primates*, evolved in the early *Paleocene Epoch* (**58-66 million years ago**). They were roughly *similar* to *squirrels* and *tree shrews* in *size* and *appearance*. The existing, very *fragmentary* fossil evidence (mostly from North Africa), suggests that they were adapted to an *arboreal* way of life in warm, moist climates. They probably were equipped with relatively good *eyesight* as well as *hands* and feet, with *pads* and *claws* for *climbing*. These *primate-like* mammals **plesadapiformes** will have to remain *shadowy* creatures for us until more data becomes available.

The *primate-like* mammals do not seem to have played an *important* role in the general *transformation* of terrestrial animal life immediately following the massive global e*xtinctions* of *plants* and *animals* that occurred about **65,500,000** years ago. The most d*ramatic* changes were brought about by the *emergence* of large grazing and browsing mammals with tough hoofs, grinding teeth, and digestive tracts *specialized* for the processing of *grass, leaves* and other fibrous *plant* materials. The evolution of these h*erbivore* mammals provided the *opportunity* for the evolution of the *carnivorous* mammals *specialized* to *eat* them. These new *hunters* and *scavengers* included *dogs, cats*, and *bears*. *Adaptive radiation* was resulting in the *rapid* evolution of *new species* to fill expanding ecological *niches*, or food-getting opportunities: *new* animals, most of which were *placental* mammals. With the exception of *bats*, none of them reached *Australia* and *New Guinea.* But the fact that there were **no people** around yet to **take them there** o*bviously* helps explain the reason why they **weren't there**. *South America* had also drifted away from *Africa* and was not connected to *North America* after **60,000,000 years ago**. *South America* reconnected with *North America,* with *placental* mammals streaming in for the *first time,* resulting in the *extinction* of most of the existing m*arsupials* there.

Early Prosimians

The first *primates* evolved by **55 million years ago**, or a bit *earlier*, near the beginning of the *Eocene Epoch, fossils* of which, have been found in *North America, Europe*, and *Asia*. They looked different than the modern *primates,* but were still somewhat *squirrel- like* in *size* and *appearance*; they had grasping *hands* and *feet* that were becoming increasingly more efficient in *manipulating* objects and *climbing* trees. The *position* of their *eyes* indicated that they were developing more effective *stereoscopic* vision.

The beginning of the *Eocene Epoch*, **55.8 million years ago**, coincides with the appearance of *early forms* of the placental *mammal orders, some of which,* still exist to this day. *Placental* mammals with larger bodies and brains additionally began to appear in the *fossil record*. Paul Falkowski has suggested that this was probably due to the a*mount* of *oxygen* in the earth's *atmosphere*, which more or less, doubled about **50 million years ago**. Since larger animals have relatively fewer *capillaries* for the d*istribution* of *oxygen* in the *cells* of their bodies, they need to breathe air that is *more* oxygenated, especially so for their brains, which require even more oxygen. *Pregnant* placement mammals, additionally, must transmit a substantial portion of the *oxygen* to their blood in their *fetuses*.

Among the new *Eocene* mammals were *primate species* that *somewhat* resemble modern, such as *lemurs*, and possibly *tarsiers*. This was the epoch of *maximum a***daptive radiation**. There were at least **60** *genera* of them that were mostly in *two families* – the *Adaptive* (similar to **lemurs**, and **lorises**) and the *Omomyidae* (possibly like g**alogos** and **tarsiers**) That was nearly a four times greater prosimian diversity than exists today. Eocene prosimians lived in *most* of the world: North America, Europe, and Asia – it was during this time that they also reached the island of *Madagascar* – in the Indian Ocean – near the coast of southeastern Africa. Their great diversity has been attributed to a lack of competition, until the subsequent arrival of monkeys and apes.

Tupaiidae (order parsimian)

Major evolutionary *changes* were beginning to *appear* in *some* of Eocene *prosimians* that were to *foreshadow* primate species *yet* to *appear*. Their *brains* and *eyes* were becoming *larger*, while their *snouts* were, conversely, getting *smaller*. At the *base* of the s*kull*, there was a *hole* through which the *spinal cord* passed. This *opening* was the f*oramen magnum* (literally the "large hole or opening" in Latin). The *position* of the foramen magnum was a strong *indicator* of the *angle* of the *spinal column* to the head and subsequently whether the *body* is habitually *horizontal* (as in a *horse*) or vertical (as a in a *monkey*). During this time, the foramen magnum in *some* species was starting to m*ove* from the *back* of the *skull* and towards the *center*. This suggests that they were beginning to hold their bodies *erect* during such activities as *hopping* and *sitting*, similar to present-day *lemurs, galagos, and tarsiers*.

By the end of the *Eocene Epoch*, many of the *prosimian species* had become *extinct*. This may be connected to *cooler* temperatures, along with the *appearance* of the *first monkeys* during the *transition* to the *next* geologic *epoch*, the *Oligocene* (about **34 million years ago**).

Tupaia (tree shrew)

A Stunning Diversity in the Tree Tops

It is in the **Early Paleocene** that we first encounter animals in the fossil record which *shows* strong *links* to our own *order*, the **primates**. Surprisingly, **primate-like** forms are best known from *continents* that are today, *apart* from their *southernmost corners*, not inhabited by *primates* other than *human beings*. This unusual *geographic* occurrence can be *explained* by the *warm,* even *subtropical* conditions that existed *far* into the *north* during *Paleocene* times. Favorable conditions also allowed other *exotic* animals like *crocodiles* to *thrive* at *high* latitudes. Yet the *importance* and *diversity* of *primate-like* mammals in the *Paleocene* faunas of *Europe* and *North America* is *remarkable*.

During the *Paleocene* most *primates-like mammals* belonged to a *group* called **Plesiadapiformes**. The *Plesiadappiformes* have traditionally been regarded as *archaic members* of the order **Primates**. Although *Plesiadapiformess* were *similar* to *modern* primates, in a number of *characteristics* of their *skeleton*, they were, nevertheless, still on a much *lower* evolutionary *level*, comparable perhaps to the *living* **tree-shrews**.

Modern primates are *unique* among *mammals* in their *adaptation* to *life* in the trees. Their capabilities of *grasping* and *leaping* allow *rapid locomotion* in this environment, which is, in turn, related to the large *brain-size* they have *developed*. As far as we know, *plesiadapiformes* also spent *most* of their *time* in the *trees*. They *lacked* adaptations for *fast leaping,* however, as we *see* them in *modern* primates, along with *not* being able to *move quickly* through the trees. Their *brains* were relatively still *very small* compared to *modern* primates. *Plesiadapiformes*, on the other hand, soon acquired *traits* that were *unusual* for *later* primates, *especially* enlarged *incisors* that were *superficially similar* to those of *rodents*. This suggests that the *plesiadapiformes* were not *direct ancestors* of *modern* primates, but rather a *branch* that split *off* from the *mainline* of *primate evolution* (from today's point of view) at an *early* date. This picture is, however, *incomplete*, complicated by *other* still *existing* orders of *mammals* that must be *close* to **primates** on the evolutionary *tree*: The **tree-shrews** (order Scandentia), the **colugos** or "**flying lemurs**" (order Dermopters), and perhaps the **bats** (order Chiroptera), although the latter are only *distantly* related, according to recent *molecule* studies.

Where *plesiadapiformes* fit in between all of these is still *hotly debated,* Many scientists, however, are today *reluctant* to call these *archaic* forms *primates*, and *regard* them instead as *members* of a separate order *Plesiadapiformes*. The most *primitive* known *plesiadapiformes*, and one of the *earliest*, is the *famous* **Purgatorius**, an animal *comparable* to a small **rat** in size. Only the remains of the *teeth* and *jaws* are known for *Purgatorius*.

Purgatorius (order plesiadapiformes)

Their *structure* is not *far* from the *pattern* that we would expect in the common ancestor of all p*lacental* mammals, presumably a small generalized **insectivore**, Yet the *teeth* of *Purgatorius* show some *traits* that are *characteristic* of both *plesiadapiformes* and *true* primates. Their *design suggests* an animal that *fed* not only on *insects* but also on some kind of *plant material*, probably *fruits*.

Restoration often *show Purgatorius* scurrying under the *feet* of **Late Cretaceous** dinosaurs like **Tyrannosaurus** and **Triceratops**. Actually one species, *Purgatorius* are from the **Early Paleocene** of **North America**. Only a *single tooth* has ever been *found*, together with the *remains* of *dinosaurs*. This *tooth* could be a *contamination* from *Paleocene rock* that was *mixed* by *mistake* with *Cretaceous* material when *collectors* sieved the sediments. Even if no *mixing* happened during *collecting*, it is possible that something *similar* occurred nearly **65 million years ago**. The *deposits* producing the single *Purgatorius* tooth were probably laid down during **Early Paleocene** time by a river cutting *deeply* into *Cretaceous* sediments, *mixing* the *fresh* remains of *Paleocene* mammals with *older* fossils, such as dinosaur *teeth*. Although unknown *primate-like mammals* may have *existed* before the *end* of the *Mesozoic*, we have *no* convincing evidence of them.

If we want to know more about *early* primate-like mammals, then *jaws* can tell us – we have to take a look later, more *specialized* forms. *By far*, the *best known* p*lesiadapiformes* are the **Plesiadapidae**, one of the most *successful* families of *Paleocene* mammals.

Plesiadapid (order plesiadapiformes)

Plesiadapids were *chipmonk-to marmot-sized* animals that had superficially *rodent- like* dentition, with a pair of robust, enlarged *incisors* that were succeeded by a long gap w*ithout* teeth (diastema) in later *forms*. Whereas the lower *incisors* were relatively simply b*uilt*, forming a kind of *scoop* together, the *upper* ones were more *complicated*, with t*hree* separate *cusps* in front and another one behind, a *configuration* that has been c*ompared* to *mittens*. Unlike the enlarged *incisors* of *rodents*, the front *teeth* of p*lesiadapids* had neither a *self-sharpening*, cutting edge; nor were they *ever-growing*, so they were probably utilized for one or several purposes *apart* from *gnawing*. In advanced p*lesiadapids* the cheek *teeth* were becoming *flattened* with the *enamel* increasingly c*renulated*. This suggests that these animals *subsisted* mainly on a *vegetable diet*, most likely *leafs* and *fruits*.

Skull and *skeleton* of **plesiadapids** were most *completely* known in the **Late Paleocene** *Plesiadapis*. This *genus* probably *arose* in *North America* and colonized *Europe* on a land-bridge via *Greenland*. Thanks to the *abundance* of the *genus* and its r*apid* evolution, species of *Plesiadapis* play an important role in the zoning of **Late Paleocene** continental sediments and in the *correlation* of *faunas* on *both sides* of the *Atlantic*. Two remarkable skeletons of *Plesiadapis*, one of them *nearly complete*, have been found in lake deposits at *Menat, France*.

Although the preservation of the *hard parts* (bone-matter) is *poor*, these *Plesiadapis* skeletons still show the *remains* of *skin* and *hair* as a carbonaceous *film* – something u*nique* among *Paleocene* mammals. Details of the *bones* are better *preserved* in the f*ossils* from *Carnay*, also in *France*, where *Plesiadapis* was one of the most *common* mammals. The rodent-like skull of *Plesiadapis* is relatively *broad* and *flat*, with a *long snout* with *orbits* still *directed* to the *side*. This is *unlike* the *forward facing* eyes of m*odern* primates that enable *three-dimensional vision*.

Despite the fact that *Plesiadapis*' braincase was *small* relative to *today's* primate s*tandards*, it was nevertheless *larger* than its contemporary *hoofed* animals. *Plesiadapis*, for instance, had mobile *limbs* that terminated in strongly curved *claws*, sporting also a long *bushy tail*, which was beautifully *preserved* in the *Menat skeletons*. The way of life of *Plesiadapis* had been the *focus* of much *debate* in the *past*. *Climbing habits* were usually *looked for* in *relatives* of *primates* – but, however, *tree-dwelling* animals, were c*onversely*, *rarely* found in the *same* large numbers as those of the *ground-dwelling-* t*ype* creatures. Accordingly, based on this fact, *paleoanthropologists* have *concluded* that *Plesiadapis* were *ground-dwelling* animals, such as today's *marmots* and *squirrels*.

More recent *investigations* have *confirmed*, however, that the *skeleton* of *Plesiadapis* is that of an *adept climber* that can be best *compared* to tree-climbing *squirrels* or to tree- dwelling *marsupials*, such as *possums*. It is *noteworthy* that early researchers surmised a close *relationship* between p*lesiadapids* and the bizarre *aye-aye* of

Madagascar, the only living *primate* with rodent- like *dentitions*. Later students have *rejected* such evolutionary *ties*, but are now, nevertheless, *regarded* as a *member* of the *same* primate *stock* as *lemurs* and *loris*. Yet the *aye-aye* may be ecologically *similar* to a rare *genus* of *plesiadapids*, the squirrel-sized *Chiromyoides* from the *Late Paleocene* (**58-66 million years ago**) of *Europe* and *North America*. **Chiromyoides** can be characterized as a super-robust *version* of *Plesiadapis*, with considerably shortened and deepened *jaw* and *muzzle*, and with extremely robust i*ncisors* that bear sharp cutting *edges*. Its feeding *mechanism* obviously had to withstand much *higher* forces than in other *plesiadapids*. The *jaws* and *teeth* of *Chiromoides* closely resembled those of the *aye-aye*, which was in previously times called *Chiromys* – hence the name *plesiadapid*. The *aye-aye* uses its enlarged *incisors* to *gnaw* into wood in search of *grubs*.

Chiromyoides may have occupied a similar *niche* in the *forests* of the **Late Paleocene**. This *specialization* would also explain its *rarity* in the mammal communities of that *time*. It may be of interest to *note* that a second *group* of **Early Teriary/ Paleocene (58-66 million years ago)** animals, the **Apatemyidae**, evolved very similar *adaptations*, probably even approximating the *aya-aya* more *closely* than most primitive

Chiromyoides.

A *second family* of *plesiadapiforms*, the **Carpolestidae** of *Paleocene* faunas in *North America*, although it never became as *dominant* as the p*lesiadapids*. As in the latter (plesiadapids), *carpolestids* had enlarged *incisors*, the *lower* ones simply built with the *upper* ones m*itten-shaped*. They were much s*maller*, however, *ranging* from the size of a *mouse* to that of a *rat*.

Carpodaptid (fruit stealer)

Carpolestids had developed their last *lower* premolars into enormous *blades* with serrated c*utting edges*, a condition termed *plagiaulacoid* that evolved *independently* in several groups of *mammals*, most *prominently* in the *archaic* **multiberculates**. In *carpolestids* these saw-like *teeth* worked *against* the also *enlarged* but *flattened* and *rasplike* third and fourth premolar of the *upper jaw*. This mechanism was probably used to process a vegetable *diet* with a high content of *fibers* such as *fruits*, *nuts* and succulent *shoots*, a diet no doubt *supplemented* by some *insects*. The *evolution* of this highly *specialized* adaptive complex creature can be *tracked* in the successive *North American* genera *Elphidorsius, Carodaptes* and *Carpolestes* that form the **Middle Paleocene** of *North America* is still very *similar* to primitive *plesiadapids*, but its last lower *premolar* already f*oreshadows* the *blade-like* form of advanced *carpoletids*. By **Late Paleocene** time, the diagnostic *traits* of the *family* were well *established* in *Carpodaptes*, and they are further r*efined* in **Late Paleocene** species of *Carpolestes* (“fruit stealer”), the *last* of the North American *carpolestids*.

Carpolestids had long been known from little more than *teeth* and *jaws,* but a *recently discovered* fossil nearly complete *skeleton* of *Carpoletes simpsoni* now sheds more *light* on their physical *appearance* and *biology.* **Carpolestes simpsoni** was a *small creature* with a body *weight* of only**100 grams**, and was obviously *highly adapted* to *life* in the t*rees*. Its *big toe* could be *opposed* to the *other toes*, enabling the animal to firmly *grasp* smaller *branches* with its *feet* just like it could with its *hands*. The big toe of *Carpolestes simsoni* carried a *nail*, something that *allowed* it to *develop* far better tactile *abilities* than if the *tip* of the *toe* was *covered* by a *claw*. **Nails** were previously known *only* in *true primates*. Their presence in a *plesiadapiform* further documents *emphatically* how these a*rchaic forms* were *adapting* to their *environment*, sometime in ways *pursued* also by *later primates*. Unlike the first true *primates*, however, *Carpolestes* showed *no* adaptations for leaping.

Another *family* of highly specialized *plesiadapiforms* were the **Picrodontidae**, tiny animals such as the type genus **Picrodus** that occurred with only a *few species* in the **Middle** and **Late Paleocene** of *North America*. **Picodontids** had long *knifelike* lower i*ncisors*, similar to those of other *plesiadaoiforms*, but their check teeth were highly m*odified*. The first *upper* and *lower* molars were extremely *enlarged*, and all molars form wide, shallow *basins* with heavily wrinkled *enamel*. Emphasis was *clearly* on increasing the *surface area* in order to *mash* some kind of *soft* food. More *abrasive* food would have o*bliterated* the low *crowns* of the *teeth* even during the very short *life span* that was t*ypical* of such *small* mammals. Their *specialized* molars were so *similar* to those of certain recent *bats* that early *researchers* even considered *picrodontids* as possible members of *that* order (bats), otherwise first known from the *Early Eocene* **(37-58 million years ago)**. This *resemblance* is today *regarded* as *convergent*, but it nevertheless provides important *clues* about the feeding habits of *picrodontids*. Like those bats, *picrodontids* probably fed on *nector* and *pollen*, a highly nutritious *diet* that was very easy on the *teeth*, and perhaps they *additionally* also fed on *tree sap*, along with f*ruits*. A *skull fragment* of the **picrodontid Zanycteris** displays a relatively narrow m*uzzle*. Any adaptation of *picrodontids* to their particular ecological *niche* remains *open* to *speculation*.

Even smaller *plesiadapiforms* belonged to the **Micromomidae**. The *family* includes diminutive forms from the **Late Paleocene** to **Early Eocene** of the *Rocky Mountain region* which carry such appropriate *names* as **Micromomys** and **Tinimomys** (the suffix of *–mys* meaning "*mouse*"). These animals may just have weighed only about **30 grams** – distinctly *less* than the smallest *living primate*, the *mouse-lemur* **Microcebus** of *Madagascar*. An amazingly complete *skeleton* of a *still* unnamed *micromomid* has recently been *discovered* in **Latest Paleocene** sediments. It suggests a tiny *creature* of only **20 gram** body weight, that was highly *adapted* for *climbing,* and may have been able to *suspend* itself from *trees*. Mammals of such *small body weight* are typically i*nsectivorous*, and in fact the sharp *teeth* of *Micromomys* confirm this was a committed insect *hunter*.

Other **plesiadapiforms** seem to have *harvested* the *diverse* fruits of the widespread t*ropical* to *subtropical* forests. Members of the **Paromomidae** are *characterized* by m*olars* that were *squared off* and *flattened*, forming large *basins* that would have been w*ell suited* for mashing *fruit*. Tree *sap* and *insects* may have *complemented* this *diet*. The d*ental* trends of **paromomyids** can be traced *back* to **Paromomys** from the **Middle Paleocene** of *North America*. In advanced forms such as **Ignacius** and **Phenacolemur** the central *incisors* become extremely *enlarged* and *slender*, and a large *diastema* develops *behind* them, like in *plesiadapids*. Both *Ignacius* and *Phenacolemur* were *long- lived* genera that ranged from the **Late Paleocene** to the **Middle Eocene**. Thus they were among the last *surviving plesiadapiforms* – *living fossils* in a world since *long inhabited* by **true primates**.

Paromyids played an *important* role from the **1900s,** which assumes *close* relationships between the *plesiadapiforms* and the *colugos* or "flying lemurs." The *latter* form one of the *smallest* existing orders of *mammals*, the **Dermoptera**, with just two s*pecies* from *Southeast Asia*. Colugos have a *gliding* membrane that runs from *behind* their *ears* to *up* the *tip* of their *tail* that allows them to *glide* over *long* distances. The g*liding* membrane of *colugos* even extends between the *digits* of *hand* and *foot*, which e*nables* them to *control* flight better than other *gliding* mammals.

Because of *similar* hand structures in **dermopterans** and **paromomyids**, it has been claimed that the *latter* also possessed a *gliding* membrane. This, along with other evidence, has led *some* researchers to the conclusion that **plesiadapiforms** are ancient d**ermopterans**, today's **colugos** being the *relics* of a once *diverse* order of *mammals* that forms the *sister* group of *primates*. More *complex* fossils, however, do *not* confirm the idea that **paromomyids** were gliders. Instead, *paromomyids* appear to have been s*quirrel-like* creatures that could *scamper* both in the *trees* and on the *ground.*

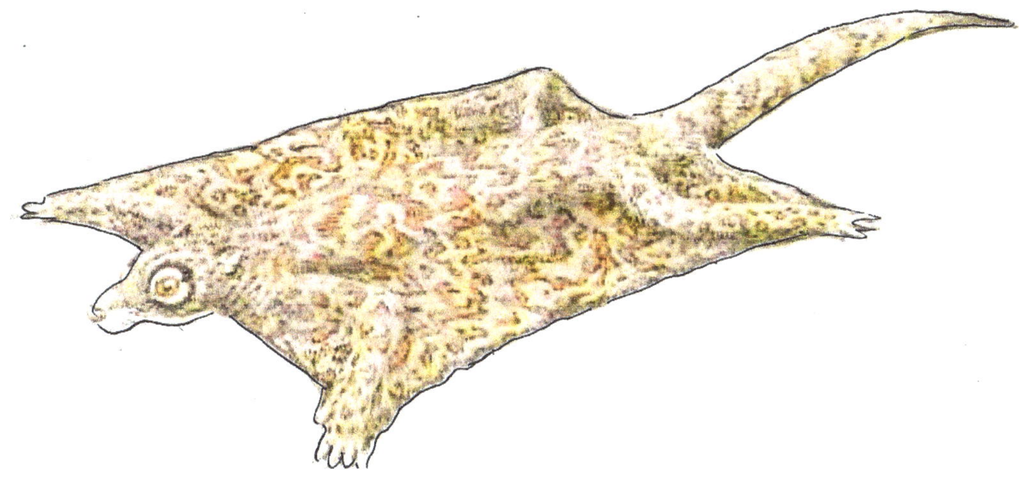

Planetetherium (a glider similar to today's "flying lemurs")

Even if no *special* relationship may *exist* between **plesiadapiforms** and **colugos**, other **Paleocene** *mammals* may prove to be early **dermopterans**. Among these are the **plagiomenids**, an *enigmatic* group that first appears in the **Middle Paleocene** of *North America. Plagiomenids* are *rare* in most *faunas*, but they can *dominate* at individual l*ocalities*, which may indicate particular *habitat* requirements. *Fossils* of the **Latest Paleocene** *plagiomenid Planetetherium,* for example, occur in *abundance* at the *coal mine* in *Montana* but are hardly known *elsewhere*. The *animals* must have found *ideal* conditions in the heavy *forested* or *swampy* environments represented by the *mine deposits*. Interestingly, during **Eocene** times some *plagiomenids* even *thrived* on *Ellesmore Island*, the *northernmost* port of *Canada*, that *lies* well *within* the *Arctic Circle*. Although *climatic* conditions where then of course were much *warmer* than *today*, the animals *living* there must still have *faced* the *winter darkness* of such *high latitudes* (at least **76** degrees North at that time).

The supposed *link* between **plagiomenids** and **colugos** is exclusively based on s*imilarities* of their *dentition*, which have long been the only *plagiomenid* fossils. The lower *incisors* of *plagiomenids*, for instance, consist of *two lobes*, perhaps an *early* stage in the *evolution* towards the specialized *comb-like* lower *incisors* that today's *colugos* use for *grooming*. Similarities of the cheek *teeth*, though somewhat vague, *suggest* also that **plagiomenids** may be an early **dermopterans**. If this is the *case*, it would *not* be surprising to find *gliding* adaptations in **Paleocene** *plagiomenids*, given the fact that other m*ammals* had *evolved* into *full-fledged* **bats** by the **Early Eocene**. This idea *cannot* be directly tested today with the available *fossils*. The link between *plagiomenids* and c*olugos* has been *weakened*, however, by a *recently* discovered *skull* of the **Latest Paleocene** to **Early Eocene** *Plagiomene*, which shows some important anatomical d*ifferences* in comparison to *dermopterans*. In face of this *conflicting* evidence, complete p*lagiomenid* skeletons are urgently *needed* to solve the *mystery* of these *intriguing* animals.

Plesiadapiforms, and perhaps **plagiomenids**, belong to a first wave of *primate-like* mammals that *flourished* in the **Paleocene**. These *archaic* forms, however, were *replaced* in the **Eocene** by **true primates**, which were called **euprimates** in technical *jargon*. The e*arliest* known *euprimates* belong to the families **Adapidae** and **Omomyidae**. Both g*roups* already *possessed* the advanced *grasping* and *leaping* adaptations, along with correlated large *brains* that characterize modern *primates*. The **adapids** were similar and perhaps related to the **lemurs** of *Madagascar* and to the **loris** of tropical *Africa* and *Asia*.

They *mainly* include larger *animals* that were presumably *active* during the *night*. **Omonyids** may include the small *creatures* with large *eyes* that suggest *activity* during the *night*.

Tarsier (order Anthropdea)

Omomyids may include the *ancestors* of the *Southeast Asian* **tarsiers**, *nocturnal hunters* of small *prey* that are great *leapers* and habitually *cling* to *branches* in a *vertical* position. **Adapids** and **omomyids** appear all of a *sudden* in the *fossil* record of *Europe* and *North America* at the *beginning* of the **Eocene**, which suggests they *immigrated* at that *time* from another *part* of the *world*, either from *Africa* or *Asia*. Alleged *euprimates* have in fact been *described* from the **Paleocene** of these *continents*. **Petrolemur** and **Decoredon** from the **Paleocene** of *China* have been *advanced* as early *members* of the **Adapidae** and **Omomydae** respectively. *Both* are *little known*, however, and it is *not* even *sure* that these animals were *related* to *primates* at all.

A more promising *candidate* was **Altialasius** from the **Late Paleocene** of *Marocco*, one of the *few* mammals that were *known* from the **Paleocene** of *Africa*. Originally described as an **omomyid**, *Altiatlasius* seemed to *support* the role of *Africa* as the *center* of primate *evolution*. Yet a recent study came to the *conclusion* that *Altialasius* was *not* an *euprimate*, but belonged instead to a mainly European group of *plesiadapiforms*, which *dispersed* into *Africa* across the ancient **Tethys** *seaway*.

We thus still *lack* conclusive *evidence* about the *early history* of **true primates** – a history which *began* without *question* during **Paleocene** times, if not earlier. We, nevertheless, have *ample* evidence of the *extraordinary* success which their *relatives*, the **plesiadapiforms** document a *varies species*.

Early Monkeys and Apes

The *Oligocene Epoch* was largely a *gap* in the primate fossil record in most parts of the world, especially true for prosimian *fossils*. Most of what we know about them came from the *Fayum* deposits in *Western* Egypt, while a *desert* today, was conversely, a t*ropical rainforest* **35 million years ago**. Other Oligocene deposits containing some fossil bones have been found in North and West Africa, the southern Arabian Peninsula, China, as well as North and South America. *Monkeys* evolved from *prosimians* during the *Oligocene* or slightly *earlier* in the *Eocene,* They were the first **species** of our s*uborder* – the **Anthopoidea**. Several general of these *early* monkeys have been **i**dentified – **Apidium** and **Aegyptoithecus** are the most *well known*. The former Apidium) was about the size of a *fat squirrel* **(2-3 pounds or 9-1.4 kg.**), while the latter (Aegyptopitheus) was the size of a large *domestic cat* (**13-20 pounds or 5.8-91kg**). Compared to the prosimians, they had *fewer* teeth, less fox-like *snouts*, *larger* brains, and increasingly more *forward-looking* eyes. These and other anatomical features suggest that the early monkeys were becoming mostly **diurnal** *fruit* and *seed* eating forest *treewellers.*

Galago (orderApidum)

New world monkeys *appeared* for the first time about **30 million years ago**. It is generally thought that they began as *isolated groups* of Old World monkeys that drifted to South America either from North America or Africa on large *clumps* of *vegetation* and *soil*. The evidence suggests that **Africa** is the most *likely* continent of *origin*. Such “floating islands” produced, as result of *powerful storms* tearing at the land, still occur in tropical regions of the world today. Due to the comparative *scarcity* of Oligocene Epoch *prosimians* in the *fossil record,* it is generally believed that the monkeys *out-competed*, and *replaced* them (prosimians), in most *environments* at that time. Supporting this *hypothesis* is the fact that modern prosimians either *live* in locations where monkeys and apes are *absent*, or are normally active *only* at *nighttime* when most of the larger, more intelligent primates are *sleeping.*

The *Oligocene* was an epoch of major *geological change,* with resulting shifts that, in the process, likely *affected* the *direction* of evolution, and thus *altered* conditions for fossil *preservation*. By the *beginning* of the Oligocene, North America and Europe drifted *apart* and became *distinct* continents. The *Great Rift Valley* system in East Africa also was formed during the Oligocene along a **1200 mile-long** volcanically *active* fault- zone, between **tectonic plates** that were moving *away* from one another. This created an e*asy* regional *migration route* for animals. By the Early Eocene Epoch **(37-58 million years ago)** *India* finally *crashed* into *Asia* and began *forcing up* the *Himalayan* chain of mountains, and the *Tibetan Plateau* beyond. During the Oligocene, the progressive *growth* of this *immense* barrier very likely *altered* continental weather patterns *significantly* by *blocking* the summer *monsoon rains*. These, along with other m*ajor* geological *events* during the Oligocene, very likely *triggered* global *climate changes*. The *cooling* and *drying* trend, with associated *expansion* of grasslands, that b*egan* in the Late Eocene Epoch, accelerated, *particularly* in the *northern hemisphere*.

The result was the general *disappearance* of *primates* from these *northern* areas. The c*limates* in most regions, nevertheless, were still *warmer* than today.

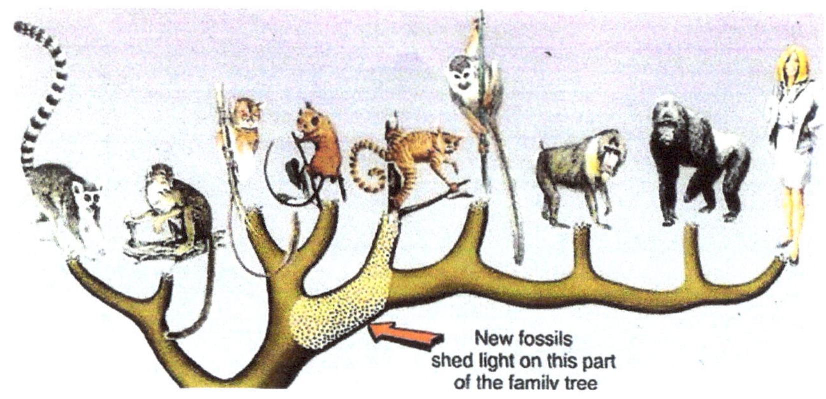

Our Family Tree

By the *middle* of the *Miocene Epoch*, the ongoing *movement* of tectonic plates created n*ew mountain chains*, which in turn, *altered* the progressive global *cooling* and *drying* trend. In addition, the cooling and drying continued, as *growing* polar ice caps *reduced* the amount of *water* in the *oceans*, causing sea levels to *drop*. This, in turn, *exposed* and r*evealed* previously *submerged* coastal lands. As a result of this continental drift, a *land connection* was *reestablished* between **Africa** and **Eurasia** that provided a *migration route* for primates along with other animals *between* these continents. Much of *East Africa* and *South Asian* tropical forests began to be replaced by sparse *woodlands* and g*rasslands* because of these *climate changes*, thus resulting in new selective pressures affecting *primate* evolution.

Primate *fossils* were *common* in the *Miocene Epoch* (**5-24 million years ago**), but not all *primates*, however, were *equally* represented in the *fossil record*. No doubt *Apes* evolved from *monkeys early* in this epoch, about **23, 000,000 million years ago**. Fossil monkeys and prosimians are comparatively *rare* from the Miocene, but *apes*, on the other hand, were *common*. It appears that apes at that time *occupied* some ecological *niches* that would be subsequently filled by *monkeys*. One of the *earliest* of monkey-to-ape t*ransitional* primates was **Proconsul**, which lived in African forests **21-14 million years ago**. Among the numerous Miocene primate species were the *ancestors* of all modern *apes* and *humans*.

Drypithecus (earliest monkey-to-ape transition)

By**14 million years ago**, the group of apes that included our *own* ancestors were apparently in the process of adapting to life on the *edges* of the expanding savannas in *Southern* Europe. They were very likely members of the genus Dryopithecus which were generally *similar* to in *appearance* to *modern* African apes. Towards the end of the Miocene, *less* hospitable *cooler* conditions in the northern hemisphere caused many primate species to become *extinct* while some, however, survived by *migrating* south into *Africa* and *South Asia*. About **8-9 million years ago**, the *descendents* of the d*ryopithecines* in Africa diverged into two lines – one that led to **gorillas** and the *other* to h**umans, chimpanzees** and **bonobos**. A *further* divergence occurred around **7-6 million years ago** which *separated* the *ancestors* of modern *chimpanzees* and *bonobos* from e*arly hominids* (human-like primates) that were our *direct descendents*.

When you *consider* the *vast* amount of time that has *passed* since life first *appeared* in the form of **single-celled prokaryotic bacteria**, more than **3 billion years ago**, it's not hard to see that *primates* are relative *newcomers* on our *planet*. The *earliest* were found in the *fossil records* dating to **50-55 million years ago**. These first **prosimians** thrived during the *Eocene Epoch*. There were no **monkeys** or **apes** to *form* any *competition* with them as *yet*. But by the time of the *transition* to the Oligocene Epoch, **about 33 million years ago**, however, *monkeys* had begun to *evolve* from the *prosimians* into the *dominant* primates, as a consequence, leaving behind many of their prosimian *ancestors* to become e*xtinct*. By the Early Miocene Epoch, about **23,000,000 years ago**, apes had evolved from *monkeys*, *displacing* them from many *environments*. In the Late Miocene, about **6 million years ago**, the evolutionary line leading to *hominids* finally became *distinct*, which included our *direct descendents!*

With a variety of feeding *strategies*, ranging from *insectivores* to generalized v*egetarians* and specialized *pollen-and-nectar-feeders*, we are just beginning to appreciate *other aspects* of their *biology*, such as *locomotion*. New *discoveries* will, hopefully, *fill* the *gap* that *separated* these *archaic forms* from *true members* of *our own order*, the **Primates**.

INTRODUCING HUMAN EVOLUTION

The *universe* is *constructed* from a *multitude* of various *materials*. It is *dynamic* in *form* and *shape* due to a multitude of various *processes* and *interactions* between these materials. To the *human*, however, in his *need* to *establish* his *place* and *purpose* in the universe, the *most* important *material* is *biological* and the most important *process* is evolution, for it is only *here* that the human can learn to *understand* himself, an understanding that is *vital* to his *survival*.

Wise men, psychiatrists, philosophers and theologians have *surmised* and *conjectured* about the *human* over the *centuries*, and *still* do, but the *truth* about *humans* may be found only *here* through *factual* knowledge. That factual knowledge lies in a *process* called **evolution**. The *human* is what *evolution made* him.

Primates

Human beings belong to the *mammalian* group known as *primates* – the scientific category that *contains* over **230 species** of *lemurs, lorises, tarsiers, monkeys of the Old and New World, and apes*. *Modern* humans, *early* humans, and *primate* species all share many *similarities* and have some important *differences*. Knowledge of these *similarities* and *differences* helps scientists to understand the *roots* of many human traits and the *significance* of each development in *human evolution*.

All *primates*, including *humans*, share at least part of a set of *common* characteristics that *distinguish* them from other *mammals*. Many of these characteristics *evolved* as *adaptations* for life in the *trees*, an *environment* in which the *earliest* primates evolved.

These *characteristics* include more reliance on *sight* and *smell*, overlapping fields of vision, allowing *stereoscopic* (three-dimensional) *sight, limbs* and *hands* adapted for *clinging* on, *leaping* from and *swinging* in the *trees*, the ability to *grasp* and *manipulate* small objects (using fingers with *nails* instead of *claws*); *large* brains in relation to *body* size; and *complex* social lives.

The scientific *classification* of primates *reflects* evolutionary *relationships* among *individual* species an *groups* of species. **Strepsirhine** (meaning "**wet nosed**") *primates* – of which the *living* representatives include *Lemurs, lorises*, along with *other* groups of species – are all commonly known as **prosimians**. *Strepsirhines* are the most *primitive* of *living* primates. They *share* all of the *basic* characteristics of *primates*, although their *brains* are neither particularly *large* nor *complex* and they have a more *elaborate* and *sensitive sense* of *smell* than do *other* primates.

The *earliest* monkeys and apes *evolved* from ancestral **haplorhine** (meaning "dry nosed") *primates*, of which the most *living* representative is the **tarsier**. Tarsiers were previously *grouped* with **prosimians**, but many scientist now recognize that *tarsiers, monkeys*, and *apes* share many traits not found in *other* primates – together make up the suborder *Anthropodea*. Anthropoid primates are divided into *New World* (North America,

South America, Central America, and Caribbean islands), and *Old World* (Africa and Eurasia) groups. The *platyrrhine* (broad-nosed) monkeys represent the first, and the second is the *catarrhine* (downward-nosed) *monkeys* and *apes*. *Humans* belong to the second group – *catarrhine*.

Apes and humans *together* make up the *super-family hominidea*, a grouping that e*mphasizes* the *close* relationship among these *species*. Scientists do *not* agree about the appropriate classification of the families *within* the super-family. *Living* hominoids are grouped into either *two* or *three* families: **Hylobatidae, Hominidae,** and sometimes **Pongidae**. *Hylobatidae* consists of the *small* or so-called *lesser* apes of Southeast Asia, commonly known as *gibbons* and *siamange*. The *Hominidae* (hominids) include *humans* and, according to some scientists, the *great apes*. For those who include *only* humans among the *Hominidae*, all of the *great apes*, including the *orangutans* of Southeast Asia, belong to the family *Pongidae*.

The *term* "hominid" traditionally referred to humans that evolved *after* the *split* between *early* humans and the other ape *lineages*. But genetic evidence, which shows c*himps* and *humans* to be more *closely* related *genetically* (evolutionary) to *one another* than to any *other* ape. This *awareness* supports *placing* all of the *great apes* and *humans* together in the family *Hominidae*. According to this *reasoning*, the evolutionary branch of Asian apes leading to *orangutans*, which *separated* from the *other* hominid branches by about **13 million years ago**, belongs to the subfamily *Poninae*. The African apes (*gorillas, chimpanzees*, and *humans*) are then classified in the subfamily called *Homininae* (or hominines). And finally, the line of *early* and *modern* humans belongs to the *tribe* (classificatory level *above* genus) *Honinini*, or *hominins*.

The classification would be *true* to the genetic *evidence*. But it tends, nevertheless, to be *confusing* when *learning* about the *subject*, as many *similar* names (*hominoid, hominid, hominine*, and *hominin*) would apply to the different *aspects* of *ape* and *human* e*volution*. In this article the term "early human" refers to *all* species of the human *family tree* since the *divergence* from a *common ancestor* with the *African* apes. Popular writing still often uses the word "hominid" to mean the *same thing*.

Humans as Primates

About *98 percent* of the *genes* in *people* and *chimpanzees* are *identical*, making *chimps* the closest *living* biological *relatives* of humans. This does *not* mean that humans *evolved* from *chimpanzees*. But it does *indicate* that *both* species *evolved* from a *common ape* ancestor. *Orangutans*, the great apes of Southeast Asia, on the other hand, *differ* genetically from *humans* to a *greater* extent, *indicating* a more *distant* evolutionary r*elationship*.

Modern humans have a number of *physical* characteristics *indicative* of an *ape ancestry*. People, for instance, have *shoulders* with a wide range of *movement*, and f*ingers* capable of strong *grasping*. In apes, these characteristics are *highly* developed as a*daptations* for *brachiation* – swinging from *branch* to *branch* in *trees*. Although humans do *not* branchiate, the general anatomy of that *earlier* adaptation still remains. Both p*eople* and *apes*, similarly have larger *brains* with greater *cognitive* abilities than do most other mammals.

Human *social life*, too, *share* similarities with that of *African* apes, along with *other* primates – such as *baboons* and *rhesus* monkeys – that live in *large* and *complex* social groups. Group *behavior* among *chimpanzees*, in particular, strongly resembles that of h*umans*. Chimps, for instance, form *long-lasting* attachments with *one another*, participate in *social bonding* activities, such as *grooming, feeding*, and *hunting*. They also form *strategic coalitions* (politics?) so as to *increase* their *status* and *power* – sound f*amiliar*?

Modern *humans fundamentally* differ from *apes,* however, in many *significant* ways.

As *intelligent* as *apes* are, for example, *our* brains are much *larger* and more *complex*, together with our intellectual *capacity* and *elaborate* forms of *culture* and co*mmunication*. This not to mention our *upright* walking *posture* and *our* ability to m*anipulate* very *small* objects, in combination with our unique throat structure (alone among primates) that gives us **speech**.

The Fossil Primates

The *origin* of the *mammalian* group *primates* is traced back to **Plesiadapiformes**, the last common *ancestors* of *strepsirines* and other mammals. *Plesiadapiformes* evolved at least **65 million years ago**. They were creatures *similar* to the modern **tree shrews**. The e*arliest* primates *evolved* about **55 million years ago**. The first *strepsirhine* primates, f*ossil* species *similar* to *lemurs* and *tarsiers*, evolved during the Eocene Epoch (about **56 to 34 million years ago**). The oldest lineage of *catarrhine* primates, from which m*onkeys, apes*, and *humans* evolved, are known between **50 to 33 million years ago**. A primate known as **Propiliopithecus** (one lineage sometimes called **Aegyptopithecus**), from the *Fayum* fossil sites of Egypt, is an archaic-looking *catanthine*, and is thought to be the common *ancestor* of what all *later* Old World *monkeys* and *apes* looked like. So *Propliopihecus* may be considered an *ancestor*, or closely *related* to a *direct ancestor* – of **humans**.

Hominoids, or members of the super-family *Hominoidea*, evolved during the *Miocene Epoch* – **24 to 5 million years ago**. Large ape species had *originated* in Africa by **23 to 22 million years ago**. Among the *oldest* known *hominoids* is a group of apes known by its genus name, *Proconsul*. Species of *Proconsul* had features that suggested a close link to the common ancestor pf *apes* and *humans*. The ape species **Proconsul heseloni** lived in dense forests of eastern Africa about **20 million years ago**. It was agile in the trees, with a flexible *backbone* and narrow chest of a *monkey*, yet capable of wide movement of the *hip* and *thumb*, as in *apes*.

Early in their *evolution*, the *large apes* underwent several *radiations*, periods when species originated and became more diverse. After *Proconsul* had *thrived* for **several million years**, a group of apes from Africa and Arabia known as the **afropithecines** *evolved* around **18 million years ago** and *diversified* into *several* species. By **15 million years ago**, apes had *migrated* to *Asia* and *Europe* over a *land-bridge* formed between the *Afrcan-Arabian* and *Eurasian* continents, which had previously been *separated*. Around this time, *two* other groups of *apes* had evolved – namely, the **kenyaoirhecines** of Africa and w*estern Asia* (first known about **15 million years ago**) and the **dryopithecines** of Europe – first known about **12 million years ago**. It is not yet *clear*, however, which of these groups of *ape species* may have given *rise* to the common *ancestor* of *African apes* and *humans*.

The First Humans: The Early Australopiths

By at least **4.4 million years ago** in *Africa*, an apelike *species* had *evolved* that had *two* important *traits*, which *distinguished* it from *other* apes: (1) small canine (eye) *teeth* (next to the *incisors*, or *front* teeth) and (2) *bipedalism* – that is the ability to *walk* on *two* legs.

Scientists commonly refer to these *earliest* human species as **australopithecines**, or a**ustralopiths** for short. The earliest *australopith* species known today belongs to the genus **Ardipithecus**. Other species belong to the genus **Australopithecus** and, by some classifications, **Paranthropus**. The name *australopithecine* translates *literally* as "southern ape," in reference to *South Africa*, where the *first* known *australopith* fossils were found. Countries in which scientists have found *australopith* fossils include *Ethiopia, Tanzania, Kenya, South Africa*, and *Chad*. Thus, *australopiths* ranged *widely* over the African continent. The *Great Rift Valley* of Eastern Africa, in particular has become f*amous* for its **australopith** finds because past *movements* in Earth's *crust* in this region were *favorable* to environments in which bones are easily *preserved* and, later, to e*xposure* of ancient deposits of fossilized bones.

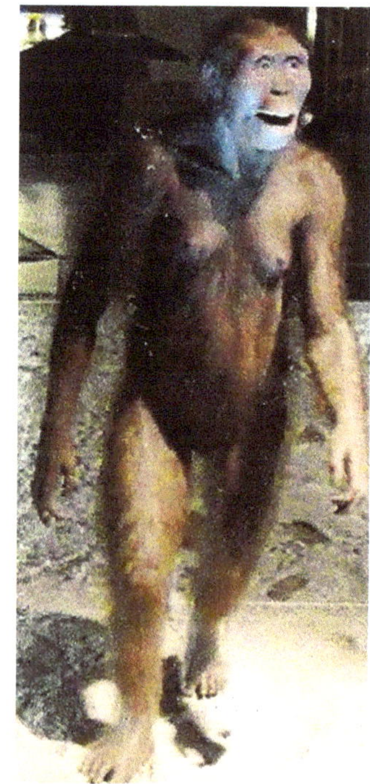
Australopithecus

There are many *ideas* about why the *early australopiths split off* from the apes, *initiating* and *setting* the course of human e*volution*. Virtually all hypothesis evoke *environmental* change as an important factor, specifically in *influencing* the evolution of *bipedalism*. Some well-established ideas about why humans first evolved include (1) the *savannah* hypothesis, (2) the *woodland-mosaic* hypothesis, and (3) the v*ariability* hypothesis. The *savannah hypothesis* argues that the *Miocene* forests of *Africa* became *sparse* and *broken up* between **5 to 8 million years ago**, due to a *cooler* and *drier* global climate. This *drying* trend led to the *separation* of an ape population in e*astern Africa* from other populations of *apes* in the more heavily forested areas of w*estern Africa*. This *eastern* population had to *adapt* to drier, open savannah environments, which *favored* the evolution of terrestrial living. Terrestrial *apes* might have formed large *social groups* so as to *improve* their *ability* to *find* and *collect* food and f*end* off *predators*. The challenges of savannah life might also have *promoted* the *rise* of tool *use*, for purposes such as *scavenging* meat from the *kills* of *predators*. These important evolutionary *changes* would have depended on increase *mental abilities* and, therefore, may have *correlated* with the *development* of *larger brains* in *early* humans.

Critics of the *savannah hypothesis* argue *against* it on *several* grounds, but p*articularly* for *two* reasons. *First*, an *early australopith* jaw *similar* to **Australopithecus afarensis** has been found in *Chad* in west central *Africa*, **2500** *kilometers* west of the *African Rift Valley*. This *find* suggests that *australopith* ranged *widely* over the African continent, and that *East Africa* may have been fully *separated* from environments *further* west. *Second*, there is growing *evidence* that *open* savannahs were not *prominent* in *Africa* until sometime after **2 million years ago**.

Criticism of the *savannah hypothesis* has *spawned* alternative ideas about early human evolution. The *woodland-mosaic* proposes that the early *australopiths* evolved in a m*osaic* of *woodland* and *grassland* that offered opportunities for *feeding* both on the g*round* and in the *trees*. Ground feeding then regular *bipedal* activity and, eventually, the evolution of anatomical features of the *hip, leg*, and *foot* that assisted this form of l*ocomotion*.

The *variability hypothesis* suggests that early *australopiths* experienced *many* changes in the environment and ended up living in a *wide range* of *habitats*, including f*orests, open-canopy woodland*, and *savannas*. In response, their *populations* became adapted to a *variety* of *surroundings*. Evidence from early *australopith* sites, In fact, s*hows* this *range* of *habitats*. So the unique appearance of their *skeletons* may have allowed them the *versatility* of living in *habitats* with either *many*, or *few* trees.

From Ape to Human

Fossils from *several* different *early australopiths* species that *lived* between **4 million to 2 million years ago**, show a *variety* of *adaptations* that *mark* the *transition* from *ape* to h*uman*. The *very early* period of this *transition*, prior to **4 million years ago**, remains p*oorly documented* in the *fossil record*, but those *fossils* that do *exist* show the most p*rimitive* combinations of *ape* and *human* features.

Fossils reveal much about he *physical build* and *activities* of early *australopiths*, but l*ittle* is known about surface physical *features*, such as the *color* and *texture* of *skin* and h*air*, or about certain *behaviors*, such as methods of obtaining *food*, or patterns of *social interaction*. For these reasons, scientists study the *living* great apes – particularly the *African apes* – to better *understand* how early *australopiths* had *similar* mental capabilities, and possibly even *social* structures.

Australopith Characteristics

Most of the distinctly *human qualities* in *australopiths* related to their *stance*. Before a*ustralopiths*, no mammal had ever *evolved* an *anatomy* for habitual *upright walking*.

African apes *move* around their *environments* in a *variety* of ways. They *use* their *arms* to c*limb* and to *swing* through the *trees* (known as *branchiation*). They *knuckle-walk* when on the *ground*, leaning on the middle parts of their *fingers*. And sometimes they *move* on t*wo legs*, as when chimpanzee *feed* on *low branches* or when gorillas show *threat* d*isplays*. The *australopith* body was devoted especially to *bipedal* walking. *Auatralopiths* also had *small canine teeth*, as compared with *long canines* found in almost all other c*atarrhine primates*.

Other characteristics of *australopiths* reflected their ape *ancestry*. Although their c*anine teeth* were *not* large, their faces stuck out *far* in *front* of the *braincase*. Their b*rains* were about the *same size* as apes today, about **390** to **550** cubic cm (**24** to **34** cubic in) but were enlarged *relative* to *body size*. Their body *weight*, which can be estimated from their *bones*, ranged from about **27** to **49** kg (**60** to **108** lb), and they stood about **1.1** to **1.5** m (**3.5** to **5** ft) tall. Their *weight* and *height* compare *closely* to those of c*himpanzees* (chimp height measured standing). Some *australopith* species had a large degree of *sexual dimorphism* – *males* were much *larger* than *females* – a *trait* also found in *gorillas, orangutans*, and some other *primates*.

Australopith also had *curved* powerful fingers and *long* thumbs with a *wide range* of movement. Apes, *in comparison*, have *longer*, very *strong*, even more *curved* fingers – which are *advantages* for *hanging* and *swinging* from *branches* – but their very *short thumbs* limit their ability to *manipulate* small *objects*. While the *fingers* were *longer* than in modern *humans*, the *australopith* finger bones were *not* so long and *curved* as to suggest *arm swinging*. It is not yet *clear* whether these *changes* in the *hand* of early *australopiths* enabled them to use *tools* in a better way than *earlier apes* or even *modern chimpanzees* today.

Bipedalism

The *anatomy* of *australopith* shows a number of adaptations for *bipedalism*. Adaptations in the *lower body* included the following: The *australopith* ilium, or *pelvic bone*, which rises *above* the *hip joint*, was much *shorter* and *broader* than it is on *apes*. This enabled the *hip muscles* to *steady* the *body* during *each* bipedal step. The *australopith* pelvis overall had evolved a more *bowl-shaped* appearance, which helped *support* the internal *organs* during upright *stance*. The *upper* legs angled *inward* from the *hip joints*, which positioned the knees to better *support* the body during *upright walking*. The legs of apes, on the other hand, are positioned almost *straight down* from the *hip*, so that when an ape walks *upright* for a short *distance*, its body *sways* from *side* to *side*. The *australopith* foot was also *reshaped*, including *shorter* and less flexible *toes*, which provided a more *rigid lever* for *pushing off* the *ground* during *each step*.

Other *adaptations* occurred *above* the *pelvis*. The *australopiths*' spine had an *S- shaped* curve, which *shortened* the overall length of the *torso*, and gave *rigidity* and *balance* when *standing*. By *contrast*, apes have a relatively *straight spine*. The *australopith* skull also had an important adaptation related to *bipedalism*. The opening at the *bottom* of the *skull*, known as the *foramen magnum*, where the spinal cord *attached* to the *brain*, was more *forward* than it is in *apes*. This *position* set the head in *balance* over the *upright spine*.

Australopiths clearly walked *upright* on the *ground*, but *paleoantropologists* debate about whether the *earliest* humans also spent at least *some* of their *time* in the *trees*. Certain *physical* features indicate that they spent at least *some* of their *time* in the trees. Such features *include* their *curved* and elongated *fingers* and *arms*.

Explaining Bipedalism

Many *different* explanations have been offered to account for the evolution of *upright walking*. Some of the ideas include: (1) *freeing* the *hands*, which was *advantageous* for carrying *food* or *tools*; (2) improved *vision*, especially to *see* over *tall grass*; (3) *reducing* the body's *exposure* to the *hot sun*, which allowed better *cooling* during the *day* in an *open* landscape; (4) *hunting* or *weapon* use, which was *easier* with *upright* posture, and (5) *feeding* from *bushes* and low *branches*, which was *easier* when *standing* and *moving upright* between *closely* spaced *bushes*.

Although *none* of these *hypothesis* has overwhelming *support*, a recent *study* of chimpanzees favors the *last one*. Chimps *move* on *two legs* most often when *feeding* on the *ground* from *bushes* and low *branches*. Chimps today are *not*, however, very *good* at *walking* in this way over *long distances*. As the distance between *trees* or *groves* of trees became wider during *drier* periods *bipedal* behavior in *pre-human* populations may have become more *frequent*. Accordingly, a more effective *bipedal gait* was favored not as an adaptation to *savannah* living but rather as a way of *crossing* less *favored* areas of *open* terrain. An ability to climb *trees* continued to be *important*. This *idea* may *currently* be the best *explanation* for the *unique* adaptation of the early *australopiths*: a *combination* of long, powerful *arms*, slightly elongated *legs*, and lower *limbs* for upright *walking* over long *distances* on the *ground*.

Small Canine Teeth

Compared with *apes*, humans have very *small* canine teeth. *Apes*, particularly *males*, have *thick, projecting, sharp* canines that they use for *displays* of *aggression* and as *weapons* to defend themselves. By **4 million years ago**, *australopiths* had *developed* the *human* characteristic of having *smaller, flatter* canines. Canine *reduction* might have *related* to an *increase* in *social* cooperation among *humans* and an accompanying *decrease* in the need for *males* to make aggressive displays.

HOMO HABILIS – The First Homo – 2.2-1-6 million years old

Findings

Man has been a *tribal animal* since he first walked *erect*, more than **four million years ago**. With the *impediment* of being *bipedal*, he could not *out-climb* nor out-run his p*redators*. Only through *tribal cooperation* could he *hold* his *predators* at bay.

For **two million years**, the only *hominid* was a *herd/tribal animal*, primarily a *herd herbivore*. During the next *two million years* the human was a *tribal hunter/warrior*. He *still* is. All of the human's *social drives* developed long before he developed i*ntellectually*. They are, therefore, *instinctive*. Such instincts as *mother-love, compassion, cooperation, curiosity, inventiveness* and competitiveness are *ancient* and *embedded* in the human. They were all necessary for the survival of the *human* and *pre-human*. Since human *social drives* are *instinctive* (not intellectual), they can *not* be *modified* through e*ducation* (presentation of knowledge for future assimilation and use). As with all other higher animals, however, *proper behavior* my be obtained through *training* (*edict* and e*xplanation* followed by *enforcement*).

The *intellect*, the *magnitude* of which s*eparates* the *human* from all other a*nimals*, developed *slowly* over the entire **four million years** or *more* of human development. The intellect is *not* unique to the *human* – It is, however, quite developed in a number of *higher* animals. The intellect developed as a *control* over *instincts* to provide *adaptable* behavior. The human is *designed* by *nature* (evolution) to *modify* any *behavior* 2that would normally be *instinctive* to one that would provide *optimal* benefit (survivability). This process is called *self-control* or *self-discipline*, and comprises the major *difference* between the *human* and the *lower order* animals, those that apply *only* instinct to their *behavioral* decisions. Self-*discipline*, therefore, is the *measuring stick* of the human. The *more* discipline behavior (behavior determined by *intellect*) displayed by the individual, the more *human* he becomes. The *less* disciplined behavior (behavior in response to *instinct)* displayed by an individual, the *more* he becomes like the lower a*nimals* that are lacking in *intellect* and are manly *driven* by their *instincts*.

Homo Habilis

Background

The *direct lineage* from the *ancestor* of both *man* and the modern *apes* to modern *man* is not *yet* known. *Evidence*, however, is *increasingly* showing thousands of relics *fitting* the general *pattern*.

The word **hominidae** is used to describe the *total* member species of the *human* family that have *lived* since the last *common ancestor* of both *man* and *apes*. A **hominid i**s an *individual* species *within* that family. The field of science which studies the human fossil record is known as **paleoanthropology**. It is the *intersection* of the discipline of p**aleontology** (the study of *ancient* life forms) and **anthropology** (the study of *humans*). Each hominid name consists of a *genre* name (e.g., *Australopology*, *Homo*) which is always *capitalized*, and a *species* name (e.g. *africanus, erectus*) which is always in *lower case*.

Some *controversy* exists on the time of this common *ancestor* to both *ape* and *human*, but it is believed to be about **5.5 million years ago**. A *key* fossil record near this time was **Ramapithecus**, which was believed to have been an *early* hominid for many years, but is now, however, considered to be an *ancient* ape that lived *near* in our common lineage. *Ramapitecus* is now thought to be an *ancestor* of the *modern* apes.

From a *genome* viewpoint, the *difference* between *modern man* and the *modern apes* is quite *small*, about **2 percent**. From a *physical* viewpoint, the *greatest* difference, however, is in **locomotion**. The human walks **upright**, generally thought to nave come about when the ancient hominid *adapted* to the *edge* of the *forest* and *plain*, and *adapted* to a life **under** the *trees* as opposed to **in** them. Fossil evidence shows that this *bipedal* adaptation was *completed* quite *early*, perhaps as early as **four million years ago**, *long* before we *looked* like or *thought* like we do *today*. *Facial* feature changes towards the m*odern* appearance came, of course, *much* later. The *facial* characteristics of *modern* man are about **100,000 years old**. The *faces* of *earlier* hominids were much *more* apelike.

Controversy exists over whether brain *size* alone shows intellectual a*bility*, but our only *measure* of intellectual *growth* in the hominid record is **brain size**. The fossil record *evidence*, except for one notable *blip*, shows a *steady* growth of brain *size*. This can, however, be *misleading*, due to the *different* sizes of the *people*.

Rampithecus

Early man (with perhaps *three* exceptions) was quite *small* with m*ales* much *larger* than *females*. From a *cultural* viewpoint, modern man and other apes are quite *similar* in *some* respects. *Sexual* practices of modern *humans* are *similar* to the *chimpanzee* (although stoutly *denied* by some), but with *far* more h*omosexual* activity. Although homosexual *play* is common among the *apes*, a *totally* homosexual ape is *rare*. It is *estimated* that about **10%** of the *human* population is *so* oriented.

The *modern* human's *trend* toward family *dissolution* places the *human* only a *few* percentage points from that of the *chimp*. In fact, unlike *man*, a *gorilla* male *must* be p*hysically* driven away and *held* at *bay* before he will *leave* his family. A great ape will r*arely* kill another member of the *same* species. On the other hand, *music* and *art* are p*eculiarities* of the *human alone* with *no* counterpart in any *ape* society.

History of man

Species Time Period

Ardipthicus ramidus	**5 to 4 million years old**
Australopithecus anamensis	**4.2 to 3.9 million years old**
Australopithecus afarensis	**4 to 2.7 million years old**
Australopithecus aficanus	**3 to 2 million years old**
Australopithecus robustus	**2.1 to 1.6 million years old**
Homo habilis	**2.2 to 1.6. million years old**
Homo erectus	**1.8 to 0.4 million years old**
Homo Sapiens archaic	**500 to 200 thousand years old**
Homo sapiens neandertalensis	**600 to 30 thousand years old**
Homo sapiens sapiens	**200 thousand years old**

The times of *existence* of the various *hominid* shown in the chart above are *based* on *dated* fossil remains. Each species may have existed *earlier* and/or *later* than shown, but fossil proof has *not yet* been discovered. There is also *dispute* concerning many o*verlapping* species – the examples of the *same* species. The same dispute exists between **Homo habilis** and **Homo erectus**, It could well be that the *two* are c*ontinuing* forms of the *same* species. The *same* dispute exists between **Homo erectus, Homo sapiens archaic,** and **Homo sapiens sapiens**. If *all* species have been d*iscovered* and the *lineage* of *man* lies *within* them, the most *probable* lineage would include *all* but the **robust Australopithecines** and the **Neanderthal**.

The following chronology is abbreviated:

The earliest fossil hominid **Ardipithecus**, is a *recent* discovery, dated **4.4 million years ago**. The remains are *incomplete* but *enough* is *available* to suggest it was *bipedal* and about **4 feet tall**. Other fossils were also found with the **ramidus** fossil which would suggest that **ramidus** was a *forest* dweller. A *new* skeleton was *recently* discovered which is about **45%** complete, and continues to be *studied*. A *new* species, **Australopithecus anamensis**, was *named* in *1995*. It was *found* in *Allia Bay* in *Kenya*. **Anamensis** lived between **4.2 to 3.9 million years ago**. Its body showed advanced *bipedal* features, but the *skull*, however, resembled *ancient* apes. **Austra- lopithecus afarensis** lived between **3.9 to 3.0 million years ago**. It *retained* the *apelike* face with *sloping* forehead, a distinct *ridge* over the *eyes*, *flat* nose and a *chinless* lower jaw. It had a brain *capacity* of about **450 cc**, and was between **3'6"** and **5'** *tall*. It was *fully* bipedal and the *thickness* of its *bones* showed that it was *quite* strong. Its *build* (ratio of weight to height) was about the same as the *modern* human, but its *head* and *face*, however, were proportionately *much* larger.

This *larger* head with *powerful* jaws is a *feature* of all species *prior* to **Homo sapiens sapiens**.

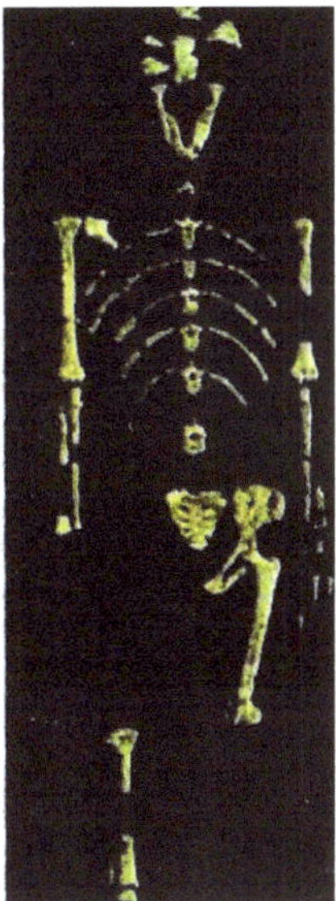

Australopithecus aranmensis

Australopithecus africanus was quite *similar* to **afarensis** and lived between **three** to **two million years ago**. It was *also* bipedal, but was *slightly* larger in *body* size. Its brain *size* was also *slightly* larger, ranging up to **500 cc**. The brain was *not* advanced enough, however, for **speech**. The *molars* were a *little* larger than *modern* humans. This hominid was a *herbivore* and ate tough, *hard-to-chew*, plants. The *shape* of the jaw was now humanlike.

Australopithecus aethiopices lived between **2.6 to 2.3 million years ago**. This *species* was probably an *ancestor* of the **robustus** and **boisei**. It ate a *rough* and *hard-to- chew* diet. It had *huge* molars and jaws and a large *sagittal crest*. A sagittal crest in a bony ridge on the *skull* extending from the *forehead* to the *back* of the head. Massive chewing muscles were *anchored* to this crest.

Australopitecus robustus lived between **two to 1.5 million years ago**. It had a body *similar* to that of **africanus**, but a *larger* and more *massive* skull and teeth. Its *huge* face was *flat*, with *no* forehead, and large brow ridges and a sagittal crest. Its brain size was up to **525cc** with *no* indication of *speech* capability.

Australopithecus boisei lived between **2.1 to 1.1 million years ago**. It was quite similar to **robustus**, but with an even *more massive* face. It had *huge* molars, the *larger* one measuring **0.9 inches** across. The brain size was about the same as **robustus**. Some authorities believe that **robustus** and **boisel** were *variants* of the *same* species.

Homo habilis was called the *handy man* because *tools* were found with his fossil remains. This species existed between **2.4 to 1.5 million years ago**. The brain size in earlier fossil specimens were about **500cc** toward the end of the species life period. The species brain shape showed evidence that **some speech** had developed. **Habilis** was about **5' tall** and weighed about **100 pounds**. Some scientists believe that h**abilis** was not a *separate* species and should be regarded either as a later **Australopithecine** or an early **Homo erectus**. It is possible that early examples in one species *group* and *later* examples in the other.

Homo Hibilis

Homo erectus lived between **1.8 million to 3000,000 years ago**. It was a *successful species* for a **million and a half years**. *Early e*xamples had a **900cc** brain size on the average. The brain, however, grew *steadily l*arger during its *reign*. Towards the end its brain size was *almost* the same size as *modern* man, at about **1200cc**. The species definitely h**ad speech**. Erectus developed *tools, weapons a*nd *fire* and learned to *cook* his food. He t*raveled* out of *Africa* into *China* and *Southeast Asia*, and developed *clothing* for *northern* climates. He turned to *hunting* for his *food*. Only his *head* and *face* differed from modern man. Like **habilis**, the face had *massive* jaws with *huge* molars, *no* chin, *thick* brow ridges, and a *long low* skull. Though *proportioned* the same, he was *sturdier* in *build* and much *stronger* than the modern human.

Homo sapiens (archaic) provides the **bridge** between **erectus** and **Homo sapiens sapiens** during the period **200,000 to 500,000 years sago**. Many *skulls* have been found with *features* intermediate *between* the *two*. The *brain* averaged about **1200cc** and **speech** was indicated, with skulls more *rounded* with *smaller* features. *Molars* and b*row ridges* were *smaller*. The skeleton showed a *stronger b*uild than modern man but w*ell proportioned*. **Homo sapiens neandertalensis** lived in *Europe* and the *Mideas b*etween **150,000 to 35,000 years ago**. **Neanderthals** *coexisted* with **Homo sapiens (archaic)** and *early* **Homo sapiens sapiens**. It is not known whether he was of the *same* species and disappeared into either the **Homo sapiens sapiens**' *gene pool*, or he may have been c*rowded* out of *existence* (killed off) by the **Homo sapiens sapiens**. Recent DNA *studies*, however, have indicated that the was entirely *different species* (rather more like *a cousin species*, sharing a *common ancestor* with **Homo sapiens sapiens**) and did *not* m*erge* into the **Homo sapiens sapiens** gene pool. Brain sizes averaged *larger t*han *modern* man at about **1450cc**, but the scan 0714 head was *shaped* differently, being l*onger* and *lower* than *modern* man. His nose was *large* and *different* from m*odern* man in *structure*. He *was massive* at about **5'6" tall,** with an skeleton *heavy* skeleton that showed a*ttachments*for *massive* muscles. He was f*ar st*onger than modern man. His *jaw* as m*assive* and had a *receding* forehead, like *erectus* **Homo sapiens sapiens** modern man) first appeared about **200,000 years ago**, and have an *average* brain size of **1350cc**. **Evolution** appears to work in *bursts* of a*ctivity*. A species may *survive* for a *long* time, even **millions of years**, with l*ittle* change, then, seemingly *overnight*, a *variant* species *springs* from it. Several such cases were *evident* among the hominid. When populations are *large*, special *drift* is very s*low*, regardless of *species*. *Evolution* works *best* when a *small* population of a species becomes *isolated* and are *suddenly* faced with *new* hazards. The environment provides e*arly* and *quick* death to quickly *weed out* deleterious m*utations* and the *small* population provides a *small* gene pool across in which *helpful* mutations may quickly *spread*.

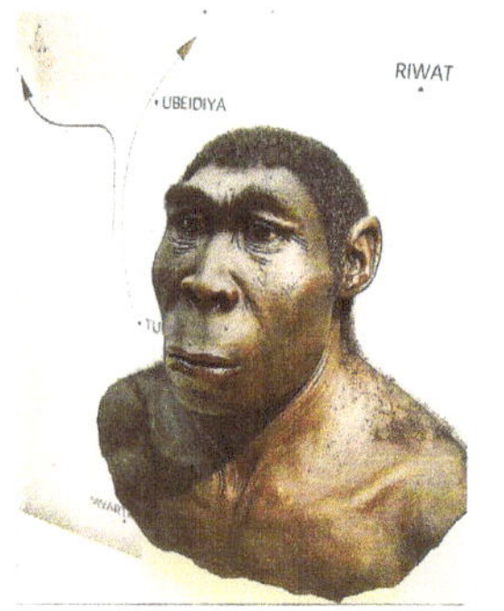

Homo erectus

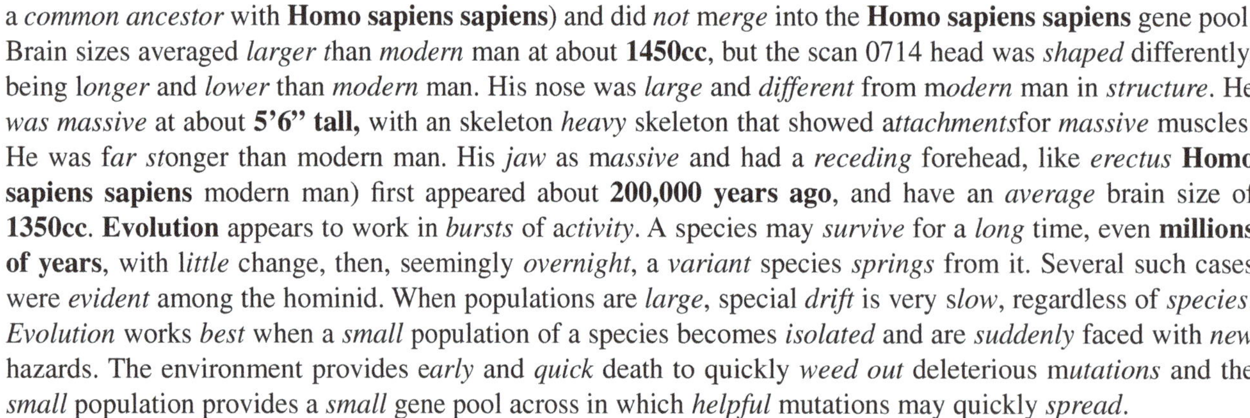

This is the manner in which the *first* hominid – the **walking ape** – appeared. Although no one knows *exactly* what happened or *where*, a small *pocket* of *primates* were somehow *isolated* in an area with *no* cats (the main primate predator) and the *food supply* was *short*, perhaps even *dwindling*.

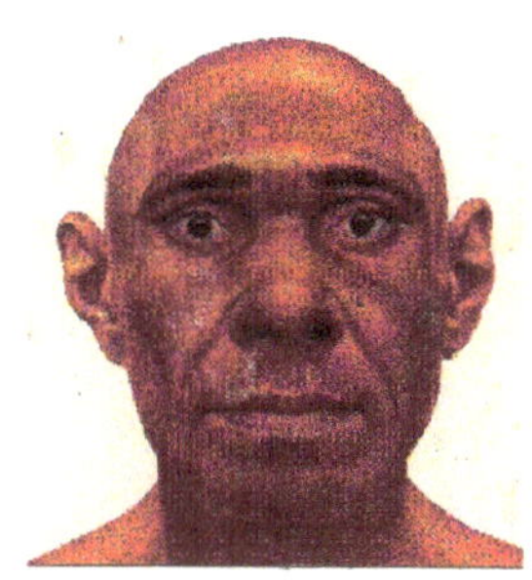

In *warmer* and *wetter* times, huge forests *abounded* across *Africa*. Both the ancient *primates* and *felines* were *widespread*. Then the climate *changed*, forests dwindled, and patches of *forests* became *isolated*, causing animal *interchange* to become quite difficult.

In most such patches, both *primate* and *feline* survived. The *shortage* of food, perhaps growing *worse* daily, *drove* some of the *primates* to the forest *floor* in search of food. There they became *food* for the *cats*. *Life* proved to be too *grim* and *short* for a new *ground dwelling* primate species to *develop*.

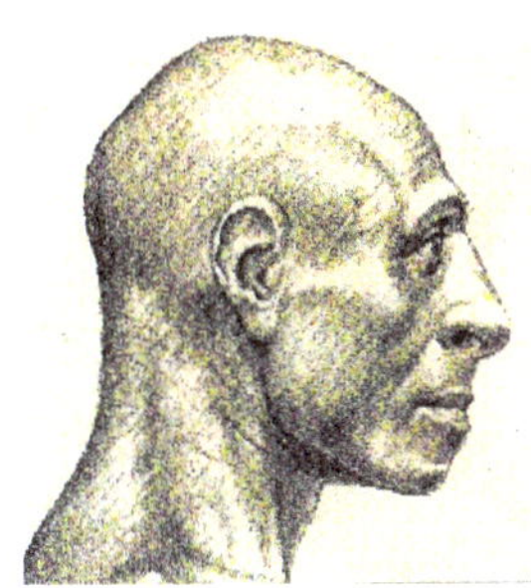

Homo sapien neanderthalensis

But somewhere there was a *unique valley*, one completely *isolated* from all the others, and something there *eliminated* the cat. Perhaps it was a *disease*, or perhaps a *famine* of *all* animal life, with the *primate* the *sole* animal survivor. In *nature*, there must always be a *large* numerical *ratio* between *food supply* and *predator*. Perhaps it was a *small* valley, too small to *support* a large enough cat *gene pool* for cat *survival*, but *large* enough to *support* bare *primate* survival. Or, more likely, the small valley was *over-harvested* by the *cat* to the point that primates *only* were *safe* high in the trees, *and survived,* while the *cat* was *starved* out of *existence*. The primate in that valley was then able to spread *safely* to the *forest floor*. The **walking ape** was **born**. The *original* primate species still *ruled* the forest *canopy*, while the *new* species, in the *absence* of *felines*, was *dominant* on the *forest floor*.

Then the climate *changed*, reopening the valley for the transit for both *primate* and *feline*. The *tree-top* primate *rejoined* his fellows and their gene pool *blended*. The *feline* was *reintroduced* to the *valley*. The *bipedal* ape on the forest floor was *introduced* to his *new* predator. If that introduction had been *sudden*, the *bipedal* ape could *not* have *survived*. Perhaps there were other valleys in which that actually happened. Luckily though, in *this* case, it was *slow,* and the **walking ape** had time to *adjust* to his *new* danger. He formed **defensive** *groups* and developed defensive *strategies*.

The *first* hominid was **Ardiphecus Ramidus**. He lived on the *forest floor*. Its close *cousin*, the *primeval* ape **Ramapithecus**, lived overhead. **Ramidus** had become a *herbivore*. **Ramapithecus** was an *omnivore*. **Ramidus** had *feet* on *one end*. **Ramapithecus** had *hands* on *both ends*. They were about the *same size* and had about the *same intelligence*. When the predator came, **Ramapithecus** escaped into the *trees*. Despite its *four hands* it could not, however, *out-climb* even the *ancient* leopard. In spite of the leopard, **Ramadus** could at least *stay* in the *forest*, being on the *open plains* was *certain death*. It was neither *fast* enough nor *strong* enough to handle the big plains *cat*, but it was, nevertheless, still *easy prey* for the leopard. The death rate, especially among the young, was *high*. A *pregnant female* had *no chance* at all. Something had to change. **Ramadus** learned how to *cooperate* in *defense* and learned how to use a *club*.

The *culture* became more *restricted* and *structured*. The *idea* of a *club* was *not* new. *Modern* chimps will use one on the *ground* in trying to *drive off* an interloper. The *chimp* does not *need* to learn how to use one *well* because it can always *take* to the high trees. Chimps will even *cooperate* in *driving* off *interlopers* by jumping *up* and *down* and *screaming*. They do not need to *learn* how to *cooperate* in *fighting*, since they can always *take* to the *trees*. **Ramidus**, however, did not have *that* choice.

Ramidus now had *two features* that kept it *out* of the *trees* in times of *danger* – its **feet** and its **club**. When the leopard came, he had *no* chance without his *club*, whether it met the cat on the *ground* or in the *tree*. Climbing a tree in a *hurry* with *two feet* that weren't *able* to *grasp* anything with a club in *one hand* while trying to *escape* a big cat would be a *harrowing* experience, at best. Its *females* and *young* had *no* chance at all without its *protection* on the ground. **Ramidus**, however, *learned* to get *shoulder-to- shoulder* with its *fellows* – club at the ready – in front of its *females* and *young*, and s*tand its ground*, regardless of *which* animal it *happened* to be. *Now* it did *not* have to *live* u*nder* the *trees*, and could live where he *pleased*. Accordingly, it *moved* on to the p**lains**.

Meanwhile, **Ramidus** was also in *deep* trouble trying to make a *living*. It was a practicing *herbivore*, but the available food, however, was *coarse* and *hard* to *chew*, *contrary* to its chewing apparatus *designed* rather to fit the *needs* of an *omnivore*, more of a *fruit eater*. The *females*, were *particularly* having *real* problems in *caring* for their y*oung* while *foraging*. The *lifestyle* was *brutal* with a *high* death rate, in line with e*volution* which *thrives* on *high* death rates.

Evolution offered **Ramidus** very few *options* – it could not afford to return to the j*ungle* since it was not *built* for it, along with *structurally* far *too slow* to *convert* to a p*redator* of the plains. It was additionally *primarily* a *vegetarian, not* equipped to t*ear* meat off his prey. Birth rate increases *would* eventually lead to major *physical changes*, but as of *now*, only *cultural* changes *were* available.

Females needed *more time* to take care of their *young*, who would *fare* much better if they were *not* exposed to the *dangers* of the *plains*. The *males* no doubt needed to take up more of the *burden*, creating a *safer heaven* for the *women* and *young*, along with p*rotection* from the *weather*. The *older* men could stand guard, while the younger men could *forage* with *clubs* in hand. The *advantage* of being bipedal *freed* their *two arms* to haul *food* back on their *return* to camp.

By the time **Australopithecus afarensis** appeared, some *structural* improvements had been made. Its head was *proportionally* larger with a much improved *eating apparatus*, with *molars* that were much *larger*. The size of the *canine teeth* had d*iminished* (*evolution* diminishes nonfunctional items). Its *jaw* was *heavier* and had *huge* chewing *muscles*. The *male* was also a little *taller* and the *female smaller* because of their *differing* roles. A slight brain size *increase* provided *improved* social interaction.

Australopithecus survived due primarily to its *harmony* with *nature* for almost **two million years**. Life, nevertheless, proved still *short*, *together* with a *high* mortality rate for the young, and *little* or *no* relief from the *constant hardship.*

Evolution had *honed* the species to *fit* the environment, which was now in *balance.* The creatures were *tough*, hard *working* and *resilient*. They had *joined* the other *plains* animals to *strike* a *balance* with *nature* that appeared *stable* (*not* fun, but at least s*urvivable*). Many *other* plain's *animals* had also reached a *stability* in their *evolutionary process*, one that *exists* to this day. If something had *not happened* to *upset* this *balance*, man would *still* be *there* today, *mingling* with the *gnus* and *wildebeests*.

Several things occurred to *spur* further development. With a *stronger* culture in place, they could survive the plains *better* than the *other* herbivores, resulting in a much higher p*opulation growth*. As *competition* for food was high, species *branched off* – **Australopithcus aethipicus** was the first, followed by **robustus** and then **boisie**. These models proved *bigger* and *tougher* competitors for the *available* food supply.

Somewhere along the line of that **million year reign** of **africanus**, *one* of them s*harpened* a stick, perhaps in an attempt to *dig roots*, and in the process, *discovered* that a s*pear* was a much more *effective weapon*. When a *club* was used *against* an *animal* other than a *fellow*, it was *immediately* available for another swing against an animal, but not, however, against another *fellow*, who could grab it on the *second* attempt, thus *losing* the advantage of *surprise*. The *aim* of the *club* was usually to *discourage*, rather than to *kill*, more effective against *animals* than its *fellows*.

A *spear* was an *offensive weapon*, having only one purpose – *to kill*. Nevertheless, s*kill* was *still* required in *its* use to avoid situations *wherein* a spear *stuck* in a adversary is f*reed* as the adversary *twists* away, thus *prolonging* the confrontation. Or in the case of a tiger, looking into its eye, barehanded, with an *out-of-reach* spear sticking out of its shoulder – not to mention, provoking the tiger's anger even further – is not exactly h*ealthy* either. On the other hand, however, *smacking* the tiger on its *head* using a *club* still leaves the *defender* armed. In *essence*, therefore, the spear works best in *sneak, stalk a*nd *kill* attacks, usual tactics employed by *hunter-killer* teams against any *animal* or *foe* on the *plains*.

Life became even more *precarious*, though a favorite *working ground* so far as e*volution* is concerned. The greatest *concerns* of these *ancestors* of man were not *animals* any more, but *other man*. When going *against* one another, *using* the *same* weapons, c*unning* was now the *deciding* factor. The *spear* proved to be the *great equalizer* over size – so *bigger* was not *better* – *smarter* was far *better*, together, *unfortunately*, with v*iciousness*. The *former* docile hominid *dweller* on the *plain* was now *transformed* into a plains *warrior*, together with a much more *complex* culture, one that now *mandated* careful *planning* along with *leadership*. This, it goes without saying, required not only more *intelligence*, but *language* also.

Homo habilis was the *transitional* ape-to-man *primate*. Starting with a **550cc** *brain*, it subsequently *grew* to a *respectable* **800cc**. **Habilis** *developed* from a brutish *ape-like* h*erd* animal to a competent *man-like one*. The *Broca's* area in its brain *developed* to a point where it *revealed* the existence of a *workable vocabulary*. **Habilis** invented the *use* of *fire* for *cooking*, *warmth* and keeping wild animals at bay, along with the *stone axe*. It may also have been responsible for the *elimination* of **robustus** and **boisei**, *whom c*oincidentally *disappeared* at this time.

Then, about **1.8 million years ago, Homo erectus (archaic) arrived**: a *mighty* warrior by this time, *skilled hunter, inventor*, far-ranging *explorer*, and *king* of *all* he s*urveyed*. It was about the *size* of a *modern* human *standing* straight; he developed a **1259cc** brain – very close to *modern* man. Along the way he *developed* many new *tools* and *weapons*, *clothing*, and *traveled* out of his homeland, *Africa*, the *first* hominid to do so. He went across *southeast Asia*, into *northern China* and south to *Java*. It was now an o*mnivore* with *meat* as its main *diet*. It *cooked* its food, thus making it *softer* and *easier* to e*at*. This *resulted* in the inevitable *evolutionary* changes, the *downgrading* of its once m*agnificent* chewing apparatus. By the end of its reign, its *molars* and *jaw* had *shrunk* to almost that of *modern* man.

The *culmination* of man's *evolution* was **Homo sapiens (archaic)**. It has been all d*ownhill* ever since. It *arrived* about **300,000 years ago**, *straight* and *tall, muscular*, h*ardened* and *practical*, with almost a *full-sized brain*, the *end-product* of **four million years of evolution**. *Humankind* was now a *veteran* of *myriad* of *deaths*, together with u*nspeakable hardships*. With a *small* enough *population* to allow *rapid mutations, h*is *gene pool* had little *variability*, but the *process* of *natural selection* kept it so *pared*, that only the *strongest*, the most *cunning* and the most *stubborn* survived.

Then came **modern man**, about **250,000 years ago**, a *more* or *less* anticlimax. From this point on, his *inventive mind* would devise *method* after *method* to *ease* his lot. He would *dispatch* his *enemies* without any *remorse*. He would *learn* how to enslave *animals* as well as his *own kind*. He would greedily *take* more than he would *give*, being responsible for more animal *extinctions* than *evolution* ever *was*, and is *now* on the verge of *extinction himself* due to *global warming!*

The Neanderthal

Concluding this story without giving *tribute* to an *enigma* in our *history* would not be proper. **The Homo sapiens neandertalensis** seems *out of context* and does *not* quite chronologically *fit* in our story. They probably came from *far northern Europe,* the probable *descendents* of an ancient **Homo erectus** tribe, a tribe that had *migrated* to that r*egion* many *hundreds* of *thousands* of years before. They had many of the physical c*haracteristics* of the modern *Eskimo*, who is *well tuned* to *arctic* living. They were s*tocky*, almost *massive*, in *build*. The *males* were about **5'6"** tall but were much *heavier* and *stronger* than *modern man*. They had large pronounced *checks*, usually associated with *cold weather* adaptation. They walked as *erectly* as *modern man*. Their *tools* paralleled the coexisting **Homo sapiens sapiens**, but it is *not* known whom *copied* who.

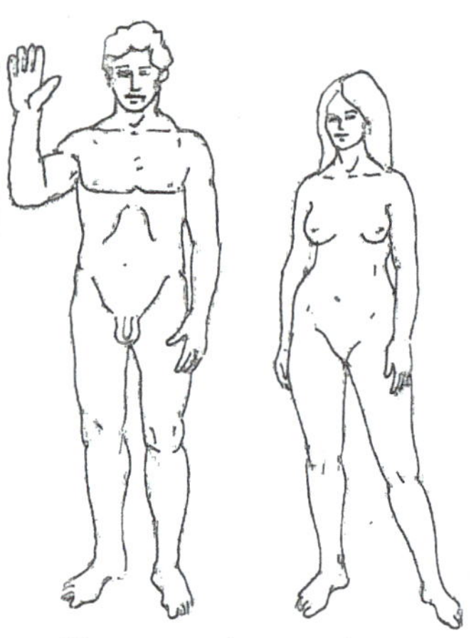

Homo sapien sapien – modern man

Although lacking a f*orehead*, they had brains that averaged **1450cc**, about **8%** larger than *modern man*. They were the *first* to *bury* their dead, *replete* with *flowers* and *artifacts*. Were they cunning *beasts*? Or were they, on the other hand, a *gentle* and i*ntelligent people?* And the b*illion dollar question – what happened to them?*

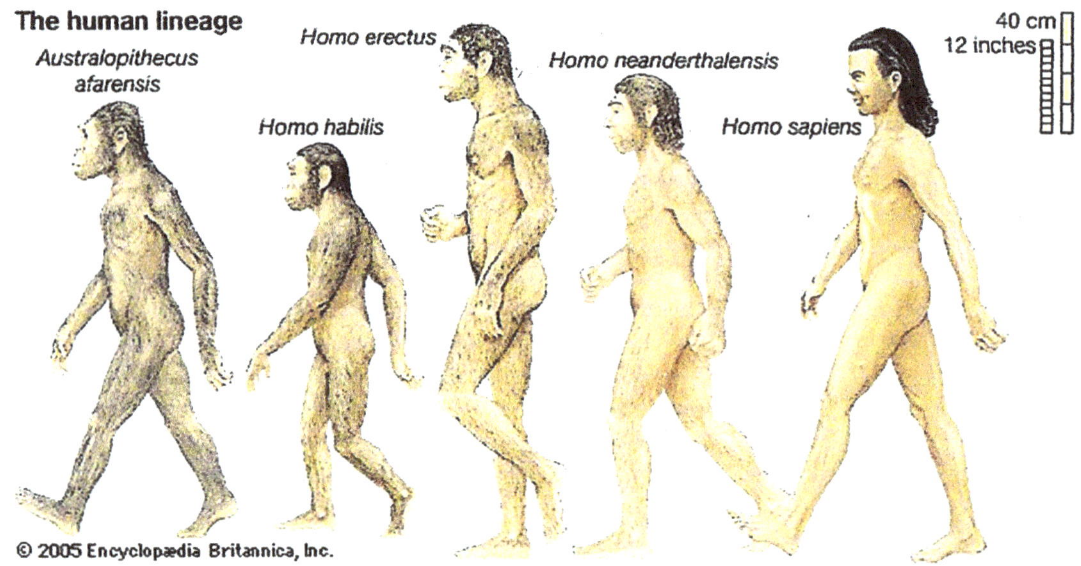

Our (Human) Lineage

Were they of the same species as us and their genes *disappeared* into a *larger pool?* Or, (most likely) did they get in the way of the early **Homo sapiens sapiens** and were simply *exterminated*? Late **e**vidence in s study of DNA, done *recently* by a *German lab*, working from *fossil remains* indicate that they were *not* assimilated into the *gene pool* of *modern man*, but instead were *separate species*, more of a *cousin* from a *common ancestor* of man.

Early Modern Homo Sapiens

All people today are classified as Homo sapiens sapiens-ie., the sapiens variety of the species of the *Homo sapiens*. The first began to appear nearly 200,000 years ago in association with technologies not unlike early Neanderthals. It is now clear that they did not come after the Neanderthals but were their contemporaries. It is likely, however, that both modern humans and Neanderthals descended from **Homo heidelbergensis**.

Compared to the Neanderthals and other late archaic Homo sapiens, modern humans generally have more delicate skeletons. Their skulls were more rounded and their brow ridges generally protrude much less. They rarely have the occipital buns found on the back of Neanderthal skulls. They also have relatively high foreheads and pointed chins.

The first fossils of early modern humans to be identified were found near the village of Les Eyzies in southwestern France. They were subsequently named the **Cro-Magnon** people. They were very similar in appearance to *modern* Europeans. Males were **5** feet to **6** feet tall (**1.6-1.8** m). That was **4-12** inches (**10-31** cm) taller than Neanderthals. The Cro-Magnons had broad, small faces with pointed chins and high foreheads. The cranial capacities were up to **1590** cm, which is relatively large even for people today.

Origins of Modern Humans

Current data suggest that *modern humans* evolved from *archaic Homo sapiens* primarily in East Africa. A **195,000 year old** fossil from the *Omo 1* site in Ethiopia shows the beginnings of the skull changes that we associate with modern people, including a rounded skull case and possibly a projecting chin. A **160,000 year old** skull from the *Herto site* in the Middle Awash area of Ethiopia also seems to be at the early stages of this transition. It had the rounded skull case but retained the large brow ridges of archaic *Homo sapiens*. Somewhat more transitional forms have been found at *Laetoli* in Tanzania dating to about **120,000 years ago**. By **115,000 years ago**, early modern humans had expanded their range to South Africa and into Southwest Asia shortly after **100,000 years ago**. They did not appear elsewhere in the Old World, evidently, until **60.000-40,000 years ago**. This was during a short temperate period in the midst of the last *ice age*.

It would seem from these dates that the location of initial modern *Homo sapiens* evolution and the direction of their dispersion from that area is obvious. That, however, is not the case. Since the early **1980s**, there have been two leading contradictory models that attempt to explain modern human evolution – the **replacement** model and the **regional continuity** model.

The *replacement* model of *Christopher Stringer* and *Peter Andrews* propose that modern humans evolved from *archaic Homo sapiens* beginning around **200,000 years ago** in Africa and then some of them migrated into the rest of the Old World replacing Neanderthals along with other late *archaic Homo sapiens* beginning around **100,000 years ago**. If this interpretation of the fossil record is correct, all people today share a relatively modern *African* ancestry. All other lines of humans that had descended from *Homo erectus* presumably became *extinct*. From this view, the regional anatomical differences that we see among humans today are recent developments – evolving in the last **40,000 years**. This hypothesis is also referred to as the "out of Africa," "Noah's ark" and "African replacement" model.

The *regional continuity* (or multi-regional evolution) model advocated by *Milford Wolpoff*, of the *University of Michigan*, proposes that modern humans evolved more or less simultaneously in all major regions of the Old World from local archaic *Homo sapiens*. Modern Chinese, for example, are seen as having evolved from Chinese archaic *Homo sapiens* and ultimately from Chinese *Homo erectus*. This would mean that the Chinese and some other peoples in the Old World have great antiquity in place.

Supporters of this model believe that the ultimate *common ancestor* of all modern people was an early *Homo erectus* in Africa who lived at least **1.8 million years ago**. It is further suggested that since then there was sufficient gene flow between Europe, Africa and Asia to prevent long-term reproductive isolation and the subsequent evolution of distinct regional species. It is argued that *intermittent* contact between people of these distant areas would have kept the human line a single species at any one time. Regional varieties, or subspecies of humans, however, are expected to have existed.

Replacement Model Arguments

There are two sources evidence supporting the *replacement* model – the *fossil record* and *DNA*. So far, the earliest finds of modern *Homo sapiens* skeletons come from Africa. They date to at least nearly **200,000 years**

ago on that continent. They appear in Southwest Asia by at least **100.000 years ago** and elsewhere in the Old World by **60,000-40.000 years ago**. Unless modern human remains dating to **200,000 years ago** or earlier are found in Europe or East Asia, it would seem that the replacement model better explains the fossil data for those regions. The DNA data supporting a *replacement*, however, are more problematical.

Beginning in the **1980s**, *Rebecca Cann,* at the *University of California*, argued that the geographic region in which modern people lived the longest should have the greatest amount of genetic diversity today. Through the comparisons of **mitochondrial DNA** *sequences* from *living people* throughout the world, she *concluded* that Africa has the *greatest genetic diversity* and, therefore, must be the *homeland* of all *modern* humans.

Assuming a specific constant rate of mutation, she further concluded that the *common ancestor* of modern people was a *woman* living about **200,000 years ago** in Africa. This supposed predecessor was dubbed "***mitochrondrial Eve***." More recent genetic research at the *University of Chicago* and *Yale University* lends support to the *replacement* model. It has shown that *variations* in the DNA of the ***Y*** *chromosome* and *chromosome* ***12*** also have the greatest *diversity* among Africans today. *John Relethford* and other critics of the r*eplacement* model, have pointed out that Africa could have had the greatest *diversity* in DNA simply because there were *more* people living there during the last several hundred thousand years. This would leave open the possibility that Africa was not necessarily the o*nly* homeland of modern humans.

Critics of the *genetic* argument for the *replacement* model also point out that the rate of mutation used for the "molecule clock" is not necessarily constant, which makes the **200,000 year** data for "*mitochrondrial Eve*" unreliable. The rate of inheritable mutations for a species can *vary* due to a number of factors including, *generation time*, the e*fficiency* of DNA repair within cells, and *varying amounts of natural environmental mutagens*. In addition, some kinds of DNA molecules are known to be subject to m*utation* than others, resulting in *faster* mutation rates. This seems to be the case with **Y** c*hromosomes* in human males.

Further criticism of the genetic argument for the *replacement* model has come from geneticists at *Oxford University*. They found that the human betaglobin gene is widely distributed in Asia but *not* in Africa. Since this gene is thought to have originated more than **200,0000 years ago**, it undercuts the claim that an African population of modern *Homo sapiens* replaced East Asian archaic *Homo sapiens* less than **60,000 years ago**.

Regional Continuity Model Arguments

Fossil evidence is also used to support the *regional continuity* model. Its advocates claim that there has been a *continuity* of some anatomical traits from *archaic Homo sapiens* in modern humans in Europe and Asia. In other words, the *Asian* and *European* physical characteristics have antiquity in these regions going back over **100,000 years**. They point to the fact that many Europeans have relatively heavy brow ridges and a high angle of their noses reminiscent of Neanderthals. Similarly, it is claimed that some Chinese facial characteristics can be seen in Asian *archaic Homo sapien* fossils from *Jinniuiushan* dating back to **200,000 years ago**. Like *Homo erectus*, East Asians today commonly have s**hovel shaped incisors** while Africans and Europeans rarely do. This supports the contention of direct genetic links between Asian *Homo erectus* and modern Asians. *Alan.*

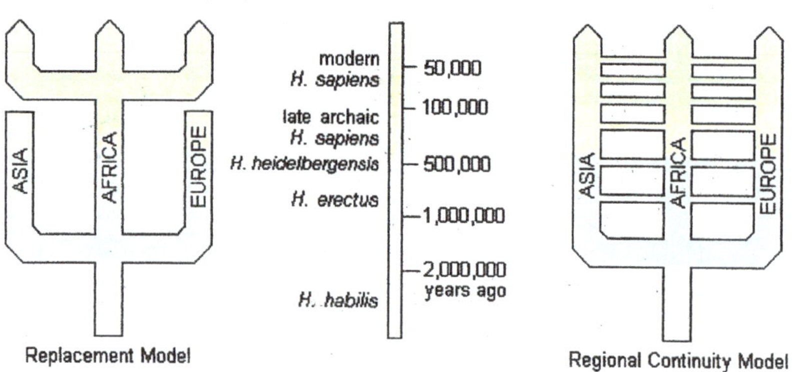

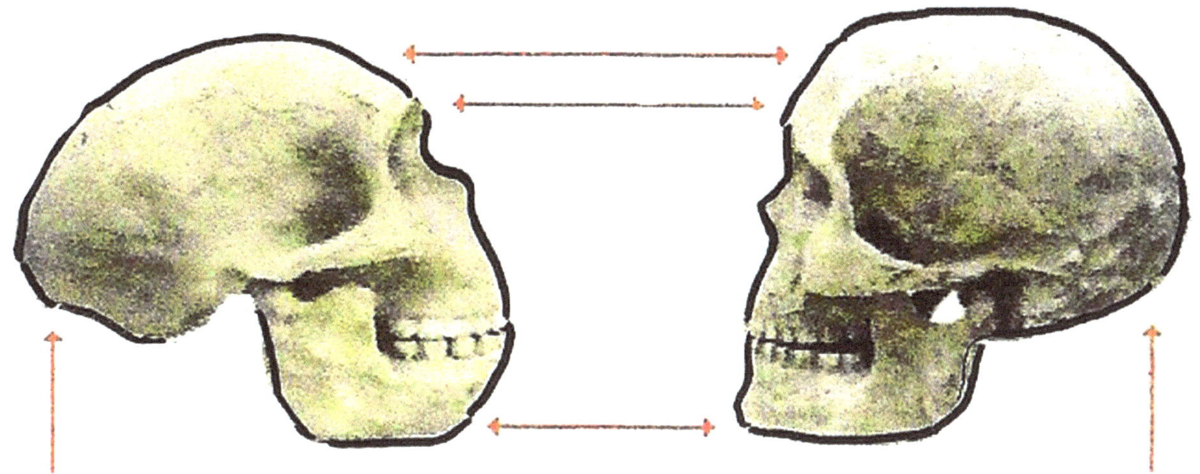

Neanderthalensis modern (us) Homo sapiens

Thorne of the *Australian National University* believes that Australian aborigines share key skeletal and dental traits with pre-modern people who inhabited Indonesia at least **100,000 years ago**. This implication is that there was no *replacement* by modern humans from Africa **60,000-40,000 years ago**. This evidence, however, does not rule out gene flow from African populations to Europe and Asia at that time and before. *David Frayer*, of the *University of Kansas*, believes that a number of European fossils from the last **50,000 years** have characteristics that are the result of archaic and modern *Homo sapiens* interbreeding.

Assimilation Model
It is apparent that both the *complete replacement* model and the *regional continuity* models have difficuity accounting for *all* of the fossil and genetic data. What has emerged is a new hypothesis known as the *assimilation* (or partial replacement) model. It takes a middle ground and incorporates both of the old models. *Gunter Brauer*, of the *University of Hamburg* in Germany, proposes that the modern humans did evolve in Africa, but when they migrated into other regions they did not simply *replace* existing human populations. Rather, they *interbred* (I agree with this argument – Author's note) to a limited degree with *archaic Homo sapiens* resulting in *hybrid* populations. In Europe, for instance, the first modern humans appear in the archeological record rather *suddenly* around **60,000 years ago**. The *abruptness* of the appearance of these *Cro-Magnon* people could be explained by their migrating into the region from Southwest Asia and possibly North Africa. They apparently shared Europe with Neanderthals for another **12,000 years**. During this long period, it is argued that *interbreeding* occurred and that the p*artially* hybridized predominately Cro-Magnon population ultimately became modern Europeans. In **2003**, a discovery was made in a Romanian cave named *Pestera cu Oase* that supports this hypothesis. It was a partial skeleton of a **15-16 year old** male *Homo sapien* who lived about **30,000 years ago** or a bit earlier. He had a *mix* of *old* and *new* anatomical features. The skull had characteristics of *both* modern and *archaic Homo sapiens*. This could be explained as the result of *interbreeding* with the Neanderthals according to *Erick Trinkaus* of *Washington University* in St. Louis. *Alan Templeton*, also of *Washington University*, reported that a computer-based analysis of 10 different human DNA sequences indicates that there has been interbreeding *between* people living in Asia,

Europe, and Africa for at least **600,000 years**. This is consistent with the hypothesis that humans *expanded* again and again out of Africa and that these emigrants *interbred* with e*xisting* populations in Asia and Europe. It is also possible that migrations were not only in one direction – people could have migrated into Africa as well. If interbreeding occurred, it may have been a *rare* event. This is supported by the fact that most skeletons of Neanderthals and Cro-Magnon people do *not* show hybrid characteristics.

Expansion Out of the Old World

The global population of modern *Homo sapiens* began to grow *rapidly* around **60,000-40,000 years ago**. It was around that time they began to migrate into regions *not* previously *occupied* by people. Their *movement* into far northern areas coincided with the end of a long cold period during the last *ice age* and had begun about **75,000 years ago**. Modern humans apparently moved into Australia for the first time between **60,000** and **40,000 years ago**. Since Australia was *not* connected to Southeast Asia by land, it is probable that the *first* Australians arrived by simple boats or rafts. Around **35,000-30,000 years ago**, human big game hunters moved into Northeastern Siberia. Some of them migrated into North America via the Bering Plain (or Beringia) **20,000-15,000 years ago** or *possibly* somewhat earlier. That intercontinental land connection appeared *between* Siberia and Alaska as a result of sea levels dropping more than **300** feet during the *final* cold period of the last ice age. Until that time, *all* human evolution had occurred in the Old World. The rate of human population growth has continued to accelerate since then.

The current world population is over **6.6 billion** and intercontinental migration and gene flow are at higher levels than ever before. A consequence of human migrations into new regions of the world has been the e*xtinction* of many animal species indigenous to these arias. By **11,000 years ago**, human hunters in the New World apparently had wiped out **135** species of mammals, including ¾ of the larger ones. Most of these extinctions apparently occurred within a **few hundred years**. It is likely that the changing climate at the end of the last ice age was also a contributing factor. The same cannot be said for the animal extinctions, however, that occurred following the arrival of *aboriginal* people in Australia, and *Polynesians* in New Zealand. In both cases, humans were *instrumental* in wiping out easily hunted species.

Large vulnerable marsupials were the main victims in Australia. Within **5,000 years** following the arrival of humans, approximately **90%** of the marsupial species larger than a domesticated cat had become extinct there. In New Zealand. It was mostly large flightless birds that were driven to *extinction* by hunters following their arrival in the **10th-13th** centuries A.D.

It is *sobering* to realize that the rate of *animal* and *plant* extinction has once again a*ccelerated* dramatically. During the last century and a half, the explosion in our human population and our rapid technological development has allowed us to move into and o*ver-exploit* most areas of our planet. This *exploitation* has usually involved cutting down forests, changing the courses of rivers, pushing wild animals and plants out of rural and urban areaas, polluting wetlands with pesticides and other man-made chemicals, and industrial-scale hunting of large land animals, whales, and fish. During the early **19th** century, there were at least **40,000,000** bison roaming the Great Plains of North America.

By the end of that century, there were only a **few hundred** remaining. They had been hunted to *near* extinction with guns. The same fate came to the African *elephant* and r*hinoceros* during the **20th** century. Likewise, commercial fishermen have depleted one species off after another during the last half century. Governments have had to step in to try to *stem* the tide of these human population effects on other species. They have, however, been only *marginally* successful. The *World Conservation Union* conservatively estimates that **7,266** animal species and **8,323** plant and lichen species are now at risk of extinction due to human caused habitat degradation. The endangered list includes **1/3** of all amphibian species, nearly ½ of the turtles and tortoises, ¼ of the mammals, **1/5** of the sharks and rays, and **1/8** of the birds. This list does not include m*illions* of *species* that are still *unknown* to science. It is likely that most of them will become extinct before they can be described and studied.

People Today

Are we *genetically* different from our *Homo sapiens* ancestors who lived **10-22.000 years ago**? The answer is almost certainly **yes**. In fact, it is very likely that the rate of evolution for our species has *continuously* accelerated since the end of the last *ice age*, roughly **10,000 years ago**. This is due mostly to the fact that our human population has grown *explosively* and *moved* into *new* kinds of *environments*, including cities, where we have been subject to *new* natural selection pressures. Our large and denser populations, for instance, have made it easier for *contagious* diseases, such as tuberculosis, small pox, and the plague, etc., to speed rapidly through communities and wreak havoc. This has exerted *strong* selection for individuals who were fortunate to have immune systems that allowed them to survive. There has also been a marked *change* in diet for most people worldwide since the last *ice age* to one that is less varied and now mainly vegetarian with heavy *dependence* on foods made from cereal grains. It is *likely* that the human species has been able to *adapt* to these an other *new* environmental pressures because it has acquired a steadily greater genetic *diversity*. A larger population naturally has *more* mutations adding variation to its gene pool simply because there are *more* people. This happens even if the mutation rate per person remains the same. The mutation rate, however, may have actually increased because we have been *exposed* to new kinds of environmental pollution that can cause additional mutations.

It is not clear what all of the *consequences* of the environmental and behavioral changes for humans have been. It does appear, however, that the average human body size has become somewhat *larger* over the last **10,000 years**, and we have acquired widespread *immunity* to the more *severe* effects of some diseases such as measles and influenza.

Finally, can we say what *direction* human evolution will take in the future? This is a fascinating question to consider but *impossible* to answer because of innumerable unknown factors. Though, it is certain that we will *continue* to *evolve* until we reach the point of *extinction*.

INTRODUCING EVOLUTIONARY PROCESS ON (US) MANKIND

Mutations are *accidents* in *reproduction*. The *only* place where such mutations can occur is in the production of the *haploid cells* (cells with a single set of *chromosomes*) in the *sperm* and *egg*, or in the joining of the two in conception. A reproduction accident anywhere *else* in the body will affect *only* the cell that suffers the accident. Such accidents will not be *added* to the *gene pool* and thus are *not* mutations. In such an accident, the sick cell is quickly *replaced* by a well one and the incident is *closed*. Yet however, when such an incident occurs in the *sperm* or *egg*, it will appear in *every* cell in the *offspring*. This mutation has then, a **50%** chance of occurring in each *grandchild*.

If the recipient of the mutation has several children, the odds are great that the mutation will join the species' *gene pool* by way of *one* or *more* children.

Natural selection then takes over and determines the *fate* of the mutation in the species' *gene pool* The *test* is *not* survivability or excellence, but rather, is in the species' p*opulation growth*. If the mutation *aids* the growth of the species' population growth, then it is *successful* and will *remain* in the gene pool. If it is not, *natural selection* will r*emove it* from the gene pool – through *hardship* and *death*.

Here are a few examples concerning **man** and **evolution** to help gain some understanding of the way evolution works. The effects shown are not necessarily *caused* by *genetics*, but *evolution* treats all *conditions* as if it were. Note that *natural selection* acts as if all genes are involved in the *success* or *failure* of an individual. Each case that e*quals* or *exceeds* the expected offspring is considered a vote for *each gene* in the g*enome*. The *mixing* of genes in *recombination* allow individual *selection* in the long period of *time*.

Effect 1: The new gene *shortens* the life to **35 years**. *Natural selection* would not see this defect as *detrimental* since the children will be *old enough* to fend for themselves by that time.

Effect 2: The *parent* has *too many* children. If so many children were born that the resulting *death* or *misery* rate *reduced* the number of the children, *evolution* would see this as *detrimental*. If *society*, however, *takes care* of the children for him they will be healthy enough to raise more children and *evolution* would than judge the *condition* as beneficial.

Effect 3: The parent *does not* take good care of his children. If society *does not* interfere by *taking care* of the children for him, the suffering children are *less likely* to raise children of their own and *evolution* would judge that the condition was *detrimental*. If society, on the other hand, *cares* for the children, *evolution* will judge that as *beneficial*.

Effect 4: The *new gene* lengthens life to **150 years**. *Evolution* will *not* see this change as b*eneficial*. Neither will it see *later* mutations that it has as *detrimental*, until the life expectancy gets so low that it *affects* child bearing and raising.

Effect 5: The man is a *murderer*. His *murder* of someone else's *children* will affect the *evolution* of the genes of *their* parents *adversely*. If the *murderer* has *sufficient* children of his own, *evolution* will not see anything *detrimental* in his lineage.

Effect 6: The man is *cruel* and *vicious* with his wife. As long as he does not *kill her* or otherwise render her *unable* to care for her children, *evolution* will see *no harm*. Even if he *kills her* and society takes over the raising of his children, *evolution* will see *no harm*.

Effect 7: The man *dies* by accident *before* he has children. *Natural selection* will see his death as *detrimental.*

Effect 8: A young lady decides *not* to *marry* and have children. *Natural selection* will see this as *detrimental.*

Effect 9: A man decides to *adopt* children instead of *having* his own. *Natural selection* will *vote* for the *genes* of the *natural parent*s of the children and vote *against* the *adoptive parents'* gene set.

Instinct and Intelligence

What is instinct? It is the *driving* force of the *behavior* of an *organism* and is directly determined by *genetic code*. Early-cell organisms, billions of years ago, developed s*ensors* to detect *light* and an *instinct* to swim towards the *light*. Others developed p*oison darts* and *sensors* to tell when another organism was near. When their sensors said that something was near, their *instinct* fired the *darts* to *obtain* a meal. With the development of sexual reproduction, the *instinct* of *sexual desire* provided the *drive* for reproduction.

The **northern pike** is a *fish* in the lakes of North America, and is a *predator*. If one is placed in a tank of water and a *smaller fish* is tossed into the tank, the pike will quickly e*at it*. If two fish of *equal size* are tossed into the pond, it will eat the *closest* one first. If two fish of *unequal size* but are placed *equally* far from the pike, it will eat the *largest* first. If the larger one is *further away*, it will eat the *larger* one *first* up to a certain distance *differential*. If the larger one is *too far* away, it will eat the small one *first*. It judges the *relative* distance of the two fish, *juggles* that with the *size* of the fish, then optimizes his chance for the *most* food. This is called *reasoning*. Yet the pike can be raised in *isolation* and it will *still* do this. This is called **instinct**. The *parameters* of the c*alculation* are *fixed* in his *genetic description*.

The **purple martin** can fly *beautifully*, directly out of the box. Flying is not easy. It requires *much skill*, but the purple martin chick does it the *first time*. Adult martins will make sure it *practices* awhile, the first time out, but it knows how to do it. That is i*nstinct*. It is *fix-wired* to fly well, *to a point*. It usually *falls* right out of the tree, the first time it *lands*. If it tries to land on a *pole*, it will usually *fall off* or *slide down*. Its *genetic code* prepared it for *flying*, those *parameters* are *fixed*, but it did not *exactly* prepare it for every possible *landing site*. It gave it *memory* and a *landing formula* for that. Watch that chick a week *later* and see a perfect landing *every time*. It has *learned* how to grab onto the *standard* surfaces. It *remembers* what it has *learned* and uses that knowledge to c*ontrol* its claws when it lands. That is *reasoning*. Yet it does not *need* another *martin* to t*each* it.

Reasoning of this type in **man** is *indistinguishable* to him from *intellectual* reasoning. The reasoning mechanism is *fixed* and is the same used in *both cases*. The only *difference* in the process is that *intellectual reasoning* follows a *learned process* (program) *stored* in a *learning* memory (RAM). The pike follows a process (program) stored in *fixed* memory (ROM). Most *reasoning* by man, which he *considers* to be *intellectual*, is *not* intellectual at all. All *cultural* (emotional) interactive *reasoning* processes are of the *fixed type*, e*mbedded* **eons ago**. Modern *data* may be *fed* into these process from *memory* or *senses*, but the process is *instinctive*. Anything involving *mother love* or *sex*, for example, will be r*easoning* following *ancient fixed* processing.

Some say that there is *no* distinction between **instinct** and **intelligence** (which includes *memory* and *reasoning*) other than in *size, relative proportions* and *complexity*, they are *dimensions* of the same *structure* and are *decreed* by the *coding* in the DNA.

Pure reasoning in the *human* is not only a *myth* but is *impossible*. His method of thinking, utilization of *memory*, and *problem solving skills*, are all *fixed* by DNA, a DNA designed by a neutral, non-reasoning **evolution** process. Man can *conceive* the idea of *pure* objective thought, but he is incapable of it. Man is a *subjective* organism.

Evolution took **millions of years** to make *sure* of it. He can talk about *objectivity* all he wants, he can foolishly believe that he is being objective, but there is no way that he can r*emove* himself from his own *instinct-reasoning* programming. Ancient *instincts* and modern *intellect* are *seamlessly intertwined* in his *brain*. The idea of *pure* objective thought is no more than another example of man's ability to *conceive* the perfect *bird* or a*nimal* trap or the advantages of moving to another valley that he has not seen yet. Man c*annot* fly to the moon. He can, however, *conceive* the idea and then build a *machine* that will take him there – which, of course – **has come to past.**

He is not ashamed in the least of *using* the machine. Still, when he considers *pure* objective thought in dealing with *personal things*, such as his *culture*, his *arrogance* compels him to believe that he can 'fly to the moon' without any *outside help*. He will become quite *irritated* with suggestions to the *contrary* (the *irritation* alone *illustrates* the d*egree* of his objectivity). Similarly, there is no reason that man cannot build a machine (perhaps even an *organic* one) which is capable of 'pure' reason. Only a machine designed for the purpose could be *completely objective*. His *biggest* problem will be in d*efining* 'pure' reason. Think of it, *intelligence without instincts*. We need it *badly*. Elect one for *president*. Put a *bunch* more on the *bench*.

The *long-term result* of **evolution** is **bare survival**. If the *organism* is in *distress*, the h*igher* death rate *removes* impediments *rapidly*. An *organism* suffering a *high* mortality r*ate* tends to become *stronger* to *match* its environment. If the organism is *stronger* than required, *evolution* will *degrade* it, again **matching** the *organism* with the *environment*. A *comfortable* organism has a *lower* death rate and so does *not* weed out *detrimental* characteristics as *quickly*. The result is a *gradual* degradation of *function* until the c*omfort-zone* is *removed*.

Back to the **pike** – if the *pike* is *successful* in its *environment*, it will *not* develop any further *intelligence*. It does not *need* it. It would be of *no value* to it. Should the environment becomes *harsher*, it will either develop *offsetting* ability or *perish*. Still, what is most likely *change*? It already knows how to *hunt*. It does not have a hand to *hold* a weapon and has *no need* to *understand* Shakespeare. Bigger *teeth*, a sleeker *body* for s*peed*, or a quicker *reaction time* would solve its problem far *better* and *quicker* than a h*igher* IQ. Look at a *pike*. It has been *gaining* these *features* for **millions of years**. *Pound* for *pound*, there is not a better **killing machine** on earth – well, maybe **man** the e*xception*!

The **martin** is in the same *fix*. It is a **born flyer**. It does not take *long*, even with a *low* IQ, to *learn* how to *fly* with its *mouth* open, *scooping* up *insects*. Only if its environment c*hanged* would it need to *learn* something *new*. Perhaps a more *agile* flying style, a d*ifferent* territory, a *bigger* mouth, or *more* broods *each year* would *solve* the problem b*etter* and *quicker* than a *higher* IQ. If it could have a *mutation* that gave it a *higher* IQ, what would be the *value*, or *use* for it? If the *new-found* intelligence is *not* required, e*volution* would *not* allow it to *last* long, as it usually does in *removing* all unused features – *applying* to *all animals!*

Why did **man** develop a **large brain**? And why, on the other hand, did the *other primates* not in turn *do so*? The answer, of course, was in his particular *environment*, how well he *matched* it, along with what the *evolutionary* alternatives were. His *upright* p*osture* was both a *blessing* and a *curse*. He found himself standing *upright* on the g*round*, but was *unable,* however, to either *outrun* or *outfight* his *predators*. It would take m*assive* changes in his *physical* structure to *improve* his situation. *Evolution* usually w*orks* in *incremental* fashion, with necessary *mutations* occurring in his *body* and *brain*.

He developed *stronger* and *better* shaped *eating* tools, but he still, nevertheless, had a *big problem* with his predators – big, fast cats, etc. In *that* respect, the *incremental* changes didn't *alleviate* the *problem* to any *great extent*. To *offset* this and help *level the ball field* for him, his *brain* grew *incrementally*, through *mutations*, and **good things** started to h**appen**. He was not only able to handle his *predator* problem *better*, but his *cultural life* improved *proportionately*. Together with *arms* and *hands* freed to carry *items* and handle w*eapons* his brain also grew *proportionately* with each *brain-size* increase.

Another *factor* in *evolution* may have brought into *play*. It is *rare* but when it *occurs* it m*ultiplies* the effect of *evolution*. It *operates* in both a *negative* and *positive* way, and a*ids* the organism in its *balance* with its *environment*, in either case. If *mutations* in a critical *area* in the DNA causes organisms *distress*, natural selection will *eliminate* the m*utation* each time it happens (through *misery* and *death*). A mutation may *occur* which p*rotects* the organism from mutations in a *critical* area. This *new mutation* will *prosper* in the *gene pool*. This is the case where one mutation *eliminates* or *reduces* another u*nfavorable* mutation *before* it happens. Its result is very *favorable* to the *organism*.

The *reverse* of that action is also *beneficial*. If a mutation in a particular area is f*avorable* (say, one that causes an increase in brain size), *natural selection* will allow that mutation to *remain* in the *gene pool* each time it occurs, if it is *needed* and *utilized*. The b*rain* will grow *incrementally* each time the *mutation* happens. Since mutations are r*andom* and *rare*, brain growth would *accordingly* also be *slow*, even if greatly *needed*.

Nevertheless, if a mutation occurred which *encouraged* such mutations so that they would happen more *often*, the rate of brain growth would be *accelerated*. This could help explain the more rapid brain growth starting at the juncture of **africanus** and **habilis**.

Still, the *coding* of an *accelerator mutation* is itself *subject* to mutation. If the *brain- power* was really making a *difference* then, this *new* mutation would be *detrimental* and be quickly *eliminated*. Yet if the organism was *comfortable*, the *accelerator* mutation would soon *disappear* from the *gene pool* since it would not affect *survivability*. The better probability happened about **100,000 years ago** when the *human population* started expanding *rapidly* (showing ability *greater* than needed). An *expanding* population is a good measure of organism comfort. *Detrimental* mutations will *accumulate* in a gene pool at such times, with *favorable* characteristics *degraded*.

Evolution, through the *liberal* application of *hardship* and *death*, had built a *strong* b*ody* and a *sound* mind by the time of the appearance of *Homo sapiens sapiens*. Both were *designed* for entirely different *environments* than *experienced* by man today. We live *longer* today for *three* reasons. *One* is **health care** and **diet**. The *second* is that our b**odies** which were *constructed* to last around a mere thirty years – due to living under brutally harsh *conditions* for countless years – now the **removal of which** *allows* us a m*uch* longer life-span. The *third* is **our culture**. We *cheat* evolution of the deaths it c*raves* to *cleanse* the *gene pool*. In the *short run* we will live *longer*. Eventually though, m*utations* will *erase* these *benefits*.

In *modern* society, survivability is no longer *dependent* on the *condition* of the *mind*. In fact, the more *successful* tend to have *fewer* children. Mutations that *distort* the f*unction* or *size* of the *brain* are no longer *removed* by natural selection from the *gene pool* The enormous *size* and still *growing* human *population* shows the spread of *adverse mutations* across the *gene pool*; but if no one dies of their effects *before* his *offspring*, alleles from adverse *mutations* will *accumulate*.

In The Brain – Size Matters

Any *mutation* must be applied to a DNA coding that already *exists*. It *cannot* be applied to *coding* that does *not* exist. Does this sound silly? No! Evolution *changes* an organism. Mutations are always *applied* to the *existing* DNA coding. Evolution makes something n**ew** out of something that *already* **exists**. If a bear becomes *distressed* in a given e*nvironment,* it does *not* sprout wings and *fly*. Instead, such things as *longer* legs or *claws* will be *tested*. Also, evolution often does *not* fix the thing that *causes* the problem, it instead *patches* the problem by doing something *unrelated*. If an organism suffers a m*utation* that *shortens* its life so that it has *difficulty* rearing its children to childbearing age, that mutation will start being *culled* from the gene pool. Before that mutation has been completely removed, another mutation may *occur* which *shortens* the *gestation* period or child *development* period. If this *shortens* the child caring requirements enough so that the shortened life is *no* longer a problem, then *both* mutations would be acceptable as *permanent* residents in the *gene pool.*

Every cell in the body can perform **any function**. *Two* copies of the *entire* genome are in *every* cell. A cell that is in the *liver* chooses to do that function. The cells in *bone* or in the *brain* choose to do those functions. When a mutation *happens*, it is either to the i*nner* function of a cell, or to the size and shape of the overall cell structure (skull, heart, etc.).

The **brain** did not start with man, of course – there were many examples of *single* cells that had *simple* versions **billions of years** before **man** appeared. *Photosynthesis* requires **light**. If a cell that depended on light drifted too low in the water, or under a land overhang that *blocked* the sun, it was in *deep* trouble. Accordingly, some developed a l*ight sensor* and a method of *swimming*. For the system to work, they developed a *central control system* that would *judge* the amount of light and if it was insufficient would turn the cell *toward* the light source and **swim** in that **direction**. It would keep *swimming* until it was *bathed* with sufficient *light*. This was *all done* within a *single* cell organism. That e*arly* cell had *memory* (what am I supposed to do?), and *reason* (Which way do I swim?)

Early animals developed *cells* that *connected* their various muscles to the *control* area. *Commands* from the *brain* drive the *muscles* through those *nerve* cells. Every *cell* in the organism *carries* all of the *information* in the DNA for the *entire* organism. Each *cell* is a u**niversal cell** and can provide *any* service in the *body* of the *organism.* Evolution c*onstructed* the *nerve* cell from the *standard* cell. It also *constructed* nerve cells that c*onnect* the various sensors (*ears, eyes, nose, skin*) to the *central area.* These *nerve cells* carry *sensor information* to the **brain**. Further cell adaptations in the central control area provide *functional* links. If the ears hear a loud *bang* they tell the *leg muscles* to *jump*! Likewise – if the *stomach* tells the *brain* is *hungry*, it *tells* the *mouth* to *open up.*

We refer to these *permanent fixed* processes as **instincts**. Still, the DNA cannot foresee all *possible* contingencies. It must allow some *leeway*. No animal is *totally* instinctive. *All animals* have *some memory*, some *reasoning* ability, and some *decision* m*aking* ability. We differ only in *degree*. The **first hominid** had all of the *neural* elements that we have today, as do the *chimp* and *poodle*. The *mutations* that *built* our brain from that *first hominid* were more about *quantity, shape*, and *organization* than in s*ubstance.*

The thing we *must* remember is that **africanus** had a **450cc** *brain size*. We have a **1350cc** *brain size*. Our *brain* is the *same one* that *crowned* **africanus** – *except*, of course, for *size*. *Evolution* just *patches* over what's there, *never* engaging in a *complete overhaul.*

Another thing – *evolution* has a ***zero*** **IQ**, and was not being *intelligent* (this the departure point between **evolutonists** and **creationists** – Author's note) when it *formed* the r*est* of our *brain*. Accordingly, it was much more *interested* in the *sex life* of our DNA.

Even *that* is not the *whole* story. **Africanus** was largely *instinctive*. Most of the *add- ons* to his brain have been *intellectual*. Those *original* instincts were *strong* and *uniform.*

Evolution saw to that. His *world* was brutally *uniform* that required his full *attention* and p*articipation*. Any *deviant* individual behavior would affect the *birthrate*. Evolution would *not* tolerate it. His *instincts* were *well maintained.*

Intelligence is *always* at *odds* with *instinct*. If the *instinct* provided *all* proper *survival* action, there would be *no need* for *intelligence*. Indeed, this is the case with all the other animals. There are literally *thousands* of *species* that *survive* quite *well* with *little* intellectual *ability*. *Intelligence* is supposed to *override* instinct to provide *action* that is more *suitable*. That is why we **got it** in the first place. By *controlling* our *instincts* we could *provide* action that *enhanced* our survivability. A little *self-discipline* provided great *survival dividends*, and *it* has obviously *worked*. Up to this point, *Man* has *conquered* the *world*! He is the *fat cat*! He is on *top* of the *heap*! *Yet now*, peak intellectual *performance* and *self-discipline* are *no longer* requirements for *survival*. Man (if you look around you) has become *self-indulgent* and has *reverted* to *satisfying* his *instincts*.

That is why today we have done a complete **180** *degrees* turn to our *basic instincts*, like **africanus,** the only *difference* is our larger **1350cc** *brain size*.

Africanus would object *loudly* to that comparison, because that *statement* is not *quite* true. We would *not last* an *hour* in his *environment*. We have *reverted* to our *instincts* – that is true – but those *instincts* are now *perverted*. Through *discipline*, man has substituted *intelligence* in place of *instinct* over a long period of time. During that *interval,* the *instincts* suffered *mutations*. Since both the original *instincts* and their *mutations* were being *overridden* by *intelligence*, the instinct *mutations* were not considered detrimental by evolution and so *accumulated* in the gene pool.

We have now *reverted* to a set of *perverted* instincts and now *cater* to these *perversions* by calling them *normal*. We *excuse* behavior now that would probably *horrify* **africanus**.

Conclusions

Conclusion 1: The mechanism for reasoning is instinctive – mechanical, fix-wired, genetically determined in function.

Argument:
The considerations of *alternative* actions and the *selection* of the most *appropriate* is not *unique* to man, along with *memory, and learned* elements of *culture* (behavior). All of these exist to *some degree* in most animals. What man calls *intelligence* is actually an *extension* of *memory*, some portions specialized, to a *tried* and *proven* reasoning mechanism, one *shared* with many *other* animals. This additional memory allowed complex *parameters* and *algorithms* to be brought into the *analysis* and *decision* function. The uniqueness in modern man's intellectual powers lies in his ability to *multiply* the *effect* of his *intellect* through external *memory* and *communications*.

A normal brain – a *gift* of millions of years of *evolution* – is able to *learn* complex reasoning processes. *Degenerative* genetic mutations to this *reasoning* mechanism are *not* subject to correction *through* training.

Conclusion 2: The mark of evolutionary success in a given set of hominid genes, is to **become a grandparent.**

Argument:
Only if the hominid result of a *genome configuration* lives *long enough* to *care* and *help* its *offspring* until they can *bear* young, will that set of genes be considered for a place in the gene pool. If there *no* offspring, the gene set is a *failure* from the standpoint of evolution. If there are offspring but they *do not*, in turn, have offspring, then the gene set is an *evolutionary failure*.

Conclusion 3: Man is not an intelligent being. He is, instead, an instinctive being **with intelligence.**

Argument:
It is obvious from the *nature* of evolution and man's *evolutionary history* that man terms intelligence was an **add-on** to an **instinctive** creature. This *new* factor *improved* man's *ability* to *survive* to become a *grandparent* by providing a wider choice in behavior and adding intellectual *control* over his *instincts*. Modern man, as it was with his *ancestors*, is *driven* by his *instincts*. He has, if he chooses to use it, *intellectual control* over his *behavior*. In ancient times the *environment* forced him to exercise this *self- discipline*. It was the *thoughtful* action that allowed him to *survive*.

Ancient man's *instincts* were his *reason* for living. His *reasoning*, on the other hand, *allowed* him to *live*.

The *removal* of the *dangers* of the *environment* from modern man (by his own *inventiveness* and *direction*) has *resulted* in *two* destructive forces, *either* of which *alone* has the *potential* to *destroy* our *species*: (1) the **halting** of gene pool *cleansing* by the *environment*, results in species *degeneration*, in *mind* and *body*, leading eventually to our species' ultimate *collapse*, and (2) the **discarding** of our *self-discipline* as a *way of life*, which with enough *time*, will *degenerate* man's *culture* (behavior) to the point of behavioral *chaos*.

Conclusion 4: Since the beginning of man, the female and the male have had **separate rules.**

Argument:
Dimorphism describes one sex *smaller* than the *other*, and is an indication of *differing rules*. In *animals* where the role is the *same*, the *size* is usually also the *same*. The *larger* and stronger *male* bore the rigors of the *defense* of the *tribe*. The smaller *female*, on the other hand, *bore* children and *maintained* the camp. During the period of the *hunter/gatherer* (essentially the **last two million years**) the *male* was the *provider*, with the *female* the family *care* giver. Since these *factors* are *no* longer *requirements* for *evolutionary* survival, *mutations* are *equalizing* the *sexes* – with equal *degeneration* – the *males* becoming more *feminine* ("girlie-men"), and the *females* more *masculine*.

Conclusion 5: The human female dictates the sexual activity

Argument:
The *hominid* female is only one in the animal kingdom who has *hidden* her fertile time so *completely* that even she is not always *aware* of when it is. Raising a *human* child is a *long-term* process. In *primitive* times, *too many* children caused a *too* high *death* rate.

Other animal *females* attract *males* when *in season*. The human *female* needed to *control* the *spacing* and *number* of her *children*. The *lack* of physical sexual *signs* required the male to always be *ready on call*. This allowed the female full intellectual *control* of her sexual activity. She used sex to *bond* the male to her so she could depend on his *help*. If he should be *killed* or *crippled*, she used it to attract a *new mate*.

A highly *monogamous* society is required for the *survival* of the human *child*, if the environment is *severe*. Not so in cases where food is *plentiful*. It appears *doomed* in *modern society*, even though there are benefits *exceeding* the need to *survive*.

Conclusion 6: Man is and always has been a tribal animal.

Argument:
Forming a *cooperative* tribe was *essential* **four million years ago**. The *hominid* was not *intellectual* at that time. The *formation* of the tribe was the result of *genetic modification*. It *was* and *is* instinctive. Hunter/gatherer societies are, by their nature, *tribal*. These *hominid* societies began **two million years ago**. After **four million years** of *tribal living*, it is safe to assume that all *mankind* is *tribal* by **instinct**.

A Basis for Morality
From both *political* (social, cultural) and *genetic* directions, the *human* is on a *collision course* with *extinction*. The worldwide cultural issues must be settled before the genetic problems can be properly approached. The cultural problem may be settled through social integration, in turn allowing a tight mobilization of the species to solve the genetic issues. Cultural integration requires a consistent, dependable and provable *moral* and *ethical* value system.

A basis for proper (moral, ethical) human behavior must be determined without reference to opinion, conjecture, spirituality, imagination, philosophy, political ideology or any other form of dogma, since these have no *real foundation*, are inconsistent, and cannot *uniformly* satisfy the needs of the human, Real (factual, scientific) knowledge is consistent and has real basis. If the human is to survive, the real knowledge uncovered in the sciences must be used as a basis for a uniform human *ethical* and *moral* behavioral system – thereby *freeing* the humans for all full attention to species survival.

The need for a consistent moral system

Within the human sub-cultures across the earth there is a chaotic mixture of personal behavior systems. All descended from ancient tribal cultures and are based on opinion, conjecture, spirituality, philosophy, imagination, political ideology, and other forms of dogma. Since the *basis* of these behavioral systems are *variable*, the resulting behaviors are also *variable*. These *differences* in behavior can be quite severe. Acceptable behavior in *one* sub-culture is often viewed with loathing by *another*. Individual movement between sub-cultures can be quite difficult, often requiring several generations to make the transition. If an individual moves into one sub-culture from another sub-culture and makes no attempt to change his behavior to match the new, he remains an *outcast*. Due to variation in language and behavioral systems, worldwide human interaction and communication suffers, often to the point of warfare. The productivity (intellectual advancement, invention) of the species is thereby diminished by the amount of intellectual assets *lost* in dealing with these variations, a loss that could be eliminated through a uniform ethical and moral behavior system.

Why is a *uniform* ethical and morl behavior system needed across the species? The answer is **two-fold**. **One** lies in current social problems which are so severe that war and terrorism may well *end* the species, if large-scale deprivation and massive infections social disease epidemics do not perform that function first. The **other** lies in a current but not yet realized *genetic* problem which is even now closing in on us on the extinction of the species.

During the **two million years** of human development as *Homo erectus*, tribes were s*mall* and *isolated*, and the entire worldwide population of the species was quite small.

Each tribe developed *genetic* and *social* differences. These differences were in both o*utward* appearance and *inner* neural mechanisms. Each tribe developed unique behaviors, dress, customs and speech. In some cases the difference was so marked as to become *racial* rather than *ethnic* differences. Each tribe was economically isolated and self-sufficient. Although some trade between tribes was probable, it was inconsequential to the survival of the tribe. Even then tribal conflict was *common* and, in fact, may have been a major factor in the *intellectual* and *social* development of the human during that period.

These tribes *still* exist, though now swollen in population and geographically overlapping. Some geographic areas contain many tribes within the same boundaries.

Geographic isolation, once so necessary for controlling conflict, has essentially disappeared with huge overlapping populations and modern transportation.

Communication has become even *more* chaotic with the advent of voice, video and digital communication via the internet and satellites. Different *languages* and *customs*, as well as other tribal behaviors, become quite troublesome. Cheek to jowl, the human struggles, often violently, to retain its individual tribal identity. As the population expands, tribal conflict can become only worse.

Another major problem is the *lack* of human goals. Evolution formed us with *no* plans in mind. As a *product* of evolution, the *human* also lives without knowing its *use* or p*urpose*. It would be helpful in developing a uniform *moral* and *ethical* behavioral system based on *real* knowledge, to first determine, if possible, the proper goals for (US) the *human species*.

What is the *end* purpose of life? Of the humans? Perhaps the answers to these questions will *never* be known, but, through a study of life itself, and the development of the human through evolution, a real process may be established. Like an *arrow* with a shaft that is **3.5 billion years old**, it points in the direction that each species must individually follow, or it, as a species, will perish. In the event that the human should become extinct, all life will likely eventually perish, for if the development of intellect by life is *not* sufficient for its survival, then the *extinction* of life itself is likely.

Species other than humans also have their developmental directions. The *cheetah* and the *antelope* are good examples. Each has been getting *faster* over the past millions of years. If, for any reason, that development should *slow* in either species, the result would be disastrous to that species. If the antelope should *gain* on the cheetah in its development of speed, the cheetah *starves*. If on the other hand, the cheetah should *gain* on the antelope in the like development of speed, the antelope may be over-hunted to *extinction*.

Each must continue developing in its *own* direction, or *perish*. Eventually one will *falter* and thus *cease* to exist.

The *race* facing the human is far different from that of the cheetah or antelope. We humans are faced with a race with the very evolution that *developed* us. The major essence of our developmental direction has been the ever-increasing application of the intellect to our behavior. Our intellect has been quite successful in nullifying environmental effects. That feature (intellect) has made us the most successful mobile species on our Mother Earth. In doing so, however, it has damaged our own evolutionary controls, resulting in a steady and rapid *degradation* of the our *intellect*. The only way this degradation can be *reversed* is by our intellectual *intervention* in the *control* of our own evolution. This, alone, is a mammoth undertaking for our species.

For the determination of human behavior based on real knowledge it is necessary to b*uild* a chain of evidence for *use* as a basis. This *evidence* must *begin* with the *first* life and *extend* through the *dawn* of Homo sapiens. It must contain the mechanisms of life and its process by which life evolves into its various forms. Having established the formation of life and its development process, there are obvious conclusions that may be drawn concerning proper human behavior. If the conclusions thereby drawn are proper, they carry the *authority* of the underlying real knowledge and may be disputed only by d*enying* that real knowledge.

Conclusion 1: In the presence of a known evolutionary direction, even in the a**bsence of known goals, the desired current behavior for a species may be determined.**

The Survival of Life

Life (DNA, the *underlying structure* of all life) began on Earth about **3.5 billion years ago**. It has *competed* with the environment and survived for all that period of time. Life survives by competing, in many cases even with *itself*. Life can be shown to be *universal*. There is, of course, only *one life* and all living things share in that life. Although the organisms (biological mechanisms, species, etc.) developed by life in competing are m*ortal* – they face natural death – life (DNA) is *immortal*, since it does *not* face natural death. In its survival, life has *developed* many functional forms (organisms, species, etc.).

Each of these forms competes to survive. That competition commonly includes: (1) competition with the **external** environment, (2) competition **between** species, and (3) competition between individuals **within** a species. Evolution is a *natural* process in life. All modern species *evolved* from other **prior** species – most now **extinct**.

Each species is a *self-replicating* group of organisms. Homo sapiens sapiens (WE, modern human) is one of these. In its present (natural) form, the evolutionary process is a s*enseless* one, without planning or goals. The opportunities for organism change (mutation accidents in the DNA replication process) is largely a matter of *chance*. The s*election* of these changes for *permanency* in the gene pool is also largely determined by c*hance*, through tending to favor those changes which *enhance* survival. Evolution is a r*eactive* system since the evolving life forms develop to survive in an environment which they do *not* control.

Within a given species, and other factors being equal, the survival of the species depends on the behavior (cultures) of that species. If it fits the current collective environment and the species does *not* become extinct, then the overall behavior of the species is the summation of the behaviors of the individuals within that species. Since the culture of a species is the *sum* of the behaviors within that culture, then the appropriateness of the individual action can be evaluated in *terms* of the characteristics of the culture.

The process of evolution is, besides being *senseless*, is also *merciless*. Among other deficient and undesirable characteristics of natural evolution, it creates species that are d*eadly* to *other* species. The resultant competition (often deadly) between species adds another dimension to the environment which *shapes* a particular species. Life makes no distinction. Since it does *not* reason and has *no* inherent direction, it merely creates life- forms. It is then the *competition* between life-forms and the *competition* between the life- forms and the physical environment that determines the set of life-forms which have the b*est* survival rates. Although not by design, nevertheless, this system *insures* that all possible physical life-forms and all possible combinations of life-forms are *tested* for survivability.

Life, in its *myriad creations*, has tried a multitude of survival mechanisms. In the case of the human species, the distinguishing factor is *intellect*. The question remains unanswered whether this is the ultimate form, the *one* which will shape all life-forms for ultimate survivability, thereby achieving immortality for life itself. Until that question is answered, each species must develop in its assigned notch and seek survivability for itself. If the human should *not* be the answer to the survival of life, care must be taken that that failure of the human does *not* harm life. In choosing between behavior alternatives, the survival of life is *permanent*. The survival of the species is *next* in importance so far as evolution is concerned. The survival of the individual is the *least* important. Since it can be assumed under present conditions, that the survival of the human species is the best chance for life itself to survive, then it is reasonable to assume that the survival of the human species must, however, take precedence in all decisions.

Since the human cannot survive without *many other* forms of life, those other species necessary for the human to survive will carry the same priority.

> *Conclusion 2*: Since the product of life is survival, normal (expected, natural) b**ehavior within a species is that which provides the optimum opportunity for species survival. Individual or group behavior which supplies less than optimum opportunity for species survival, is perverted (not normal).**

> ***Conclusion 3*: In the evolutionary process, mutations occur to individuals primarily by chance, without regard to the safety or comfort of the individual. The environment then removes those mutations which are deleterious to species survival through death and suffering to the individual. The natural process of life includes both mutations and merciless screening. The end result is the survival of the species (community) as opposed to the survival of the individual. In the natural process of life, the behavior and survival of the individual are subservient to the species' welfare.**

For the first **180,000 years** of species' existence, the modern human (Homo sapiens sapiens) was a tribal/warrior/ hunter. He *still* is. The *hominid* has been a tribal animal for the past **four million years**, and the human (Homo) for the past **2 million years**. Tribal behavior is *instinctive*. All social drives (care of children, cooperation, competition, tool and weapon creation, territorial defense, language, dress, song, art, compassion, etc.) can be shown to be *based* on *instinct*, though the final form and execution is *influenced* by i*ntellect*. All of these facets of culture are directly related and interdependent and must exist in some form in all of the higher organisms. All modern cultures are based on these social drives. The *details* of culture may *vary* from *group* to *group*, depending, of course, on a given viewpoint.

Whereas technology requires *facts*, and therefore is uniformly applied from group to group, while all cultural studies are *based* on dogma of one sort or another. Whereas in technology a *truth* is a *truth*; in culture a truth is a matter of group opinion. It (opinion) is in fact, quite variable even within groups.

Individual social behavior within a particular cultural group becomes a matter of accumulated *dogma* (*opinion, individual philosophy, conjecture, hearsay, gossip*, in *imaginations, etc.*), applied under any set of circumstances. It is no wonder that every possible behavior may be found *within* any given culture.

The question is asked: If there are necessary behavioral rules, why can't they be expressed in the same *objective* manner as in our *technology*, thus ending cultural *variability*? Such a culture would be based on *knowledge* rather than *dogma*, and go a long way towards developing a *cohesive* culture under which all could live productively – free of present prevalent on-going tribal conflicts (*war, terrorism, genocide ethnic cleansing, racial bigotry, etc.*).

It goes without saying that all species practice *deviant* behavior, but only we (humans) are the only ones with **intent** and able to **understand** it, and thus *correct* it.

***Conclusion 4*: Since the human has choice in behavior, and because of his intellect understands the consequences of perverted behavior, unlike the other species, can control these instinctive drives and thus bring his behavior in conformity with the community standards. The end result of *intellect* over *instinct* produces a more "human" individual.**

It is quite probable that life will never be a *consequence* in the universe. Chances are it will *sputter* for a while then *disappear*, possibly not even lasting until the *demise* of the earth itself. A glimmer of hope for an effect on the universe by life exists in the human intellect. The universe has *no purpose*. It *exists only*. *Intelligent design* is a *new* concept in the universe. Until it *developed* in the *human* there was *no* intellect to *endow* the universe with *purpose,* to bring about workable, uniform culture as a working environment. This is *essential* for a *creative* species. An intellectual culture, one based on real knowledge, is required to *provide* the creative atmosphere so *essential* for human invention, and thereby survival.

A Genetic Challenge

he *human* is a definable biological survival mechanism, one developed to modern form some **200,000 years ago**, in an environment that can also be defined. Its expected individual behavior may be determined from a study of that dynamic interactive system.

A culture (behavioral system) developed from this real knowledge would be an intellectual culture. All *prior* dogma must give way in case of conflict.

The *distinguishing* feature of the survival pattern developed by evolution in the human species is the *ability* to modify and control natural (instinctive) behaviors with intellectual consideration of human behavior has been very successful in solving human environmental problems. Unfortunately that very success has distorted the natural evolutionary process resulting in a *destructive* evolution that will *degenerate* the human species into *extinction* unless order is brought by human *intervention*. This intervention will require great *invention* and *dedication*. The survival and fulfillment of the human species, therefore, depends on the continued development and use of the human intellect, which, in turn, may only be achieved through proper human behavior.

The proper behavioral controls for all mobile species are reactive. Under these controls, current behavior is a direct result of current environment. These controls are called *instincts*. Superimposed over these instincts in the human, and largely in *competition* with these instincts, are various *intellectual* controls. These developed as *modifiers* to the instinctive controls because they provided more optimum survival behavior than available with the *raw* instincts. The human *substitutes* intellectual control for instinct, when determining proper action (behavior). That substitution is called '*self- discipline*' or '*self-control.*'

From the *beginning*, the human has used its intellect to make its lot easier. Rather than e*ndure* the stresses of the environment, it first invented *clothing, shelter*, and *hunting tools*. Later it invented *agriculture, manufacturing, medical care* and a *compassionate culture*. All of these subsidized deleterious mutations that would have otherwise been c*ulled* by the *environment*.

A new field in the *structure* of the human is *dawning*, one concerning human i*ntervention* in the genetic structure of the human. Surely we need not suffer the thousands of genetic defects now resident in the gene pool. A *large* portion of these defects are neural and therefore adversely offset the community culture. Conceivably this intervention could become quite extensive even to the point, in the future, of the creation of a *new* human species to *replace* our own. Modern sub-culture (group behavior systems) are based on *dogma*. Most of these will strenuously object.

Many modern philosophers on the subject predict dire results on tampering with human DNA configuration. Most objections are on *ethical* grounds. Some of these fears are quite valid. Many call for extremely *limited* or *no* intervention at all. What they all fail to see, however, is that we have *no* choice. It is as simple as that. We either will *gain* engineering control over our evolution or become *extinct*. That is predictable logic – simply put. *Timing* is critical. We do not have much of it left to bring our evolutionary process under control. Even at this writing the social structures of the world seem to be coming apart at the seams. Should we *fail* to regain engineering control over this chaotic situation we (humans) will become *history*.

The area of concern, of course, is *human behavior*, strongly influenced by the competing conflict between human *instinct* and *intellect*, the former (instinct) slowly losing ground to the latter (intellect).

Evolution degenerates characteristics not screened by the environment. If instincts are controlled by intellect, they no longer need to breed true, since the intellect will dictate proper behavior. The originally dominant social instincts will, in time, and slowly atrophy. As these instincts atrophy it becomes more urgent to bolster and this more effectively enforce community rules, to offset instinctive moral weakening, before it gets to the point of diluted law enforcement, a point where intellectual restraints no longer apply. The total collapse of these instinctual moral stops will no doubt impact severely on future human survival.

The human species has *developed* an unusual evolutionary process. The natural *process* of *evolution*, is a process that has produced a myriad of wonderful life forms, which depends primarily on 2 factors: (1) a **replication** process which at times randomly produces genetically defective offspring, and (2) a **challenging** environment that quickly s*creens out* these aberrant changes *not* beneficial to the organism, thus allowing only those offspring to survive so as to *contribute* to the *improvement* of the species.

Evolution *operates* in a merciless *random trial* and *error* process, interested mainly in s*urvival* of the species – paradoxally, using death as a tool to periodically *cleanse* the gene pool. A species which *suffers* the vicissitudes of a harsh environment will have a high birth rate to go with a like high death rate and short life-span, but its gene pool will be accordingly well maintained by evolution. On the other hand, in a *benign* environment, deleterious mutations are ejected from the gene pool thus allowing only beneficial mutations to propagate the species.

Modern *social atrophy* continues to occur because of our *diluted* instincts. The process that brought this about is not necessarily *irreversible* – we cannot, of course, again return to the good old days of our old jungle habitat *assuming* a primate configuration, *operating* basically on our instincts, where life was much less complicated. To avoid extinction, we need to *listen* to our *instincts* and apply them as a moral *barometer*, working together with our *intellect,* to *beat* the odds and turn them in our favor, to better increase our chances to survive as a species.

Darwin and Intelligent Design

Charles Darwin (1809-1682) is of course remembered for his **theory on evolution**. Much *controversy* has always, and still surrounds Darwin's theory. Questions abound. Is evolution a **3 ½ billion year old process**, creating life forms primarily at *random* but each shaped by an ever-changing and complex environment, that has *resulted* in all of the wondrous life around us? Or are all of these elements of our nature, along with the v*astness* and *majesty* of the *entire universe*, a *creation* of our imagination and empathy, one far greater than anyone can imagine? Questions are being asked. Did man really e*volve* from the *ape?* Or, is it that both *man* and *ape* are, along with other flora and fauna, creations of a *giant* intellect that *first* perceived an unfathomable spiritual need and created all *time, mass, space, light,* and *life itself* as it is today directly from *His thought*, and only a **few thousand years ago?**

Or could it be that these two concepts, though seemingly worlds apart, are merely two different sides of the *same coin*? Could it be possible that the theory of evolution is not a d*enial* of God, but description of how God created a system that in turn produced the m*iracle* of *life* and all of its manifestations? Some say that a *belief* in *evolution* will lead to a destruction of *rules* of *moral* behavior resulting in a destruction of human culture.

But a close look at evolution shows instead a need for a system of personal behavior (ethics, morality) for mankind that is *far more* strict than any religion, a need so great that if not heeded, and soon, the human will be quite likely *self-extinct*.

To begin with, the sum total of Darwin's theory is considered by the public to be "man d*escended* from the ape" and it is so *judged* in religious arguments. The fact is that in attempting to judge his theory for religious analysis purposes, they do not consider the v*ast* scientific progress in *genetics* since Darwin's time. Judging evolution on the basis of the "the man evolved from the ape" is much like *evaluating* the Sahara Desert on the basis of a *grain* of sand, or judging a large city by *looking* at a city limits sign at the edge of town.

It was *not* a new idea with Darwin, that species *developed* from *other species* based on competitive survival within the *stresses* of an *environment*, but it was Darwin, after a lifetime of study and thought, who wrote "*Origin of the Species*" and presented his theoretical argument to the public. As a young *passenger* on an anchored British frigate, the *Beagle*, off the *Galapagos Islands,* Darwin was able to *study* the Islands, a small and isolated ecosystem, and the life-forms that existed there. He had quickly realized that if evolution as a theory could *exist* as truth, it would be *evident* here. Although working under a rather constricted time-line, *mandated* by the *Beagle's* sailing schedule, his ship, he had a *few* days *only* to study there, in which to *produce* an unbiased work, based entirely on evidence *obtained* strictly by *observation* during his *short* study of a tiny island fragment of the earth that was quite young from a *geological* standpoint. It was a very *limited* study of a very *limited* scope of the *condition* of life that had existed only a very short time (geologically speaking). Any intelligent person knows that it is dangerous to *extrapolate* from the *specific* to the *general*. Because one may see the moon rise shortly after sunset on a particular night, it is foolish to believe and then try to convince everyone else in the world to also believe that it will *do so* at the same time every night, everywhere in the world and forever. Yet that is *exactly* what Darwin claimed.

Darwin's theory, some say and perhaps rightly so, is a *flimsy* description of a hole- ridden idea full of inconsistencies, guesswork and opinion. Fossils offered in evidence have doubtful application and may be explained in other ways that are as plausible as those offered in favor of evolution. Scientists admit that fossil dating is at best only a*pproximate* and they disagree among themselves in determining where a particular fossil fits in the evolutionary chain. Is a particular specimen an *early* form of man? Or is it a*nother* form of life that was created and then fell by the wayside? Or is it an example of a very disease ridden and *distorted* skeleton from some hapless ancestor of ours.

It is apparent that there *other* issues at stake, issues of great importance that must be considered. For all of the time in man's history, his behavior has depended to a large extent on his *culture* and his culture has been *based* largely on *religious concepts*. Human behavior was *determined* by these religious concepts. People were *judged* using religious concepts as a basis. If these concepts were *not* true, then the entire culture could collapse.

People *feared* the consequences when something was discovered, or thought to be discovered, that appeared to be *contrary* to religious thinking and dogma.

But one must realize, even with these objections raised, they have *no* connection with the *truth* of the theory. But the *implications* are vast. The *integrity* of a number of religions could become *suspect*. And the cultures *based* on these religions could be shaken to the core. If *accepted*, would a culture shock result? Would mankind be better off *without* such knowledge even if it is *true*? Or would culture *chaos* be the *product* of such *knowledge*? Still, as humans, creatures that are *intellectual* and *disciplined*, not ones that *react* through *instinct* alone, we must *know* the truth and *face* the consequences by basing our actions on *truth*. The lid to Pandora's box (in this case filled with the facts of evidence and its consequences) is open and its *contents* are scattered to the winds, never to be *ignored* again. So, the human must *look* deeper to find truth, one way or another. It is the *nature* of mankind.

To properly *view* the current disagreement between those who do and do not accept Darwin, and to *establish* a method for approaching a solution to such an argument, it would be wise to *look* at similar arguments in the past. There have been many but we will look only at *two* of these for background.

The Sphertical Earth Theory
Ancient Egyptians had *long* known the earth was round. By stationing several observers along a long (hundreds of miles) measured north-south path in the Nile valley, all observing the angle to the sun at high noon on the same day, they were able to determine that the earth was not only *not* flat but had a *curved* surface, one that if logically extended would *also* measure the circumference of the earth.

Aristotle (about **350 BC**) noted the shape of the earth by *observing* its shadow on the face of the moon *during* a lunar eclipse. But the Christian church held firm in its b*elief* that the earth was flat. Not only did they hold that it was flat, they *judged* as c*riminal* all who *held* a different view. Almost **1500 years** after Aristotle, they jailed Roger Bacon (**1214-1294**) for **10 years** for his views about the *spherical* earth. The *Catholic Encyclopedia* of the period lists about Roger Bacon, "He was an author full of heresies and suspected views." The Christian church *held* the flat earth views into the 16th century AD and so no one *dared* dispute it.

But the *telescope* had been *invented* and was becoming *common*. It became increasingly apparent that either the sky *was* a gross fabrication by a perverse deity else the earth was spherical. It was not long before *no* reasonable person believed in the flat earth notion although a definite proof was *yet* to be available. Then along came **Isaac Newton** (**1643-1725**) who *provided* the *mathematical* and *physical* reasoning to put the l**ast nail** in the flat earth coffin. *Only* the very *ignorant* or very *stupid* believed a *flat* earth a*fter* Newton's work.

It was *found* that scientifically determined fact was *not* detrimental to the church after all.

A description of how God did something is NOT a denial of God!
The Heliocentric Theory
Did the *sun* circle the earth? Or was it the *earth* that circled the sun? The church was adamant! The earth was *fixed* and if any circling was done it was done by everything else, and the Bible seemed to back up their position (King James version):

1 Chronicles 16.30: "The world is also established, that it can not be moved." Psalm **93:1**: "Thou has fixed the earth immovable and firm . . ."

The idea that the earth *circled* the sun, not the *reverse*, first appeared in the western world with early philosophers Pythagoras, Philolaus and Aristrchus. It lay dormant for many centuries. It *first* intruded on religious theory in Europe with the works from *Nicholas Copernicus* about **1500 AD**. Still *it* lacked *authority* being based on o*bservations* that did not constitute *proof*. Gallileo Galilei (**1564-1642**), an *early* worker with the *telescope*, went *public* with his observations and theories. The Catholic church t*hreatened* him with *excommunication* and a possible trip to the church dungeons for a lesson in *torture*. He was *forced* to publicly testify that his findings had been given him by the *devil* in order to cause God and the church trouble. He *lived* in poverty and disgrace the *remainder* of his life.

A rewrite of **Psalm 93:1** could *bring* known scientific knowledge into harmony.

Doing so would have been *gibberish* to the people of that time, however, and would have been detrimental to the growth of the church. The **Psalm**, for example, could well have been: **The new Psalm 93:1:** He *created* a *vast universe* consisting of *time, space, matter* and *energy* where all of these dimensions of reality are relative in order to provide an unlimited environmental opportunity in which His people and all other forms of life could grow. He created earth as a birth habitat for life and then provided a life generating mechanism that would allow all living things to be spontaneously created according to His plan. **He hast fixed the earth immovable and firm** *within* this magnificent s*tructure.*

Nevertheless, even though the proper **Psalm 93:1** exists *unchanged*, it has been found that scientifically determined fact was **not** *detrimental* to the church after all.

A description of how God did something is not a denial of God
Intelligent Design
So the current *disagreement* between *church* and *science* surrounds Darwin and his theory. The *same* thing that has happened *before* is *now* happening *again. Bitter* words fly *back* and *forth* while the science is busy in the laboratories making measurement after measurement while the politician is making *hay* for himself with the controversy.

To *believe* that the human is an *intelligent design* is *turning* a *blind* eye to the very basic beliefs that man is **imperfect** and therefore *needs* **religion** to guide him through **life**. Using the views of the religion is it not *more* proper to realize that man is a **work in progress**, in the *process* of evolution where *evolution* is the process *designed* by God that is *capable* of producing **perfection**? And that man *still* needs a *lot* of **development**?

Doesn't evolution *describe* that **need**? Doesn't it *provide* for a *perfection* that has *not* been **realized**? Shouldn't we *consider* the real facts that can be proven through observation and practice to be true? Isn't it *wrong* to reject *measurable* and *provable* fact merely from an *uninformed* and *reactionary* stance? Did He *endow* us with **reason** only to **confuse** us?

The fact **is** that the human *is* an *outstanding* example of a very **poor design**. The study of our DNA is in its infancy yet more than **15,000** known *defects* that cause untold misery, insanity and even death have been *catalogued*. It is a creature that walks *upright* but *has* the *spine* and sinus design of a *creature* that walks on *all fours*, resulting in much human misery especially with age. It *was* designed with the ability to be *logical* then blessed with a set of *instincts* working in *adverse* directions. It *spends* much of its time and assets in *killing* one another. It *robs, rapes* and *murders*. A peaceful country requires armies to protect itself. A large portion of our civil expenditures provide for lawyers, judges and police. *Some* will *murder* their own *unborn* babies. *Others* will do the same immediately after birth (so-called crib death???). Who Knows **for sure**? (and what about a**bortion? –** Author's note) They tend to believe in *superstitions* and *religions* that are extremely *destructive* in some cases and *counter productive* in others. They *build* huge m*onuments* to themselves. They *teach* peace and *wage* war (*Moslems, Nazis, etc.*). Even the Christian church *once* waged war, *tortured* unbelievers and took slaves. Both sexes w*aste* a sizable portion of their *time* and *assets* in the hope of enticing *illicit sex. abortion, lipstick*, and *condoms* are all big businesses. The most *advertised* medications on television are *aphrodisiacs*. The modern human *raises* its young in schools that teach emancipation from personal discipline and *relieves* them from *responsibility* by *excusing* their actions. Then we *provide* television for their recreation loaded with every imaginable social ill for entertainment. The *parents* then *abandon* the *child* except for t*ransportation* between the *school* and the *television* set.

This creature is *hardly* the design of a *vast* and *superior* intelligence – unless, of course, one should believe that the construction of His design is *still* **in progress** using a *process* which *we* call **evolution**.

The Darwin Theory

Some people tend to see (perhaps deliberately) only a *small* part of Darwin's Theory and then *ignore* the rest. That *portion* is "**man descended from the ape**," which of course *he didn't* say at all in the first place. That phrase when spoken *alone* tends to *upset* people a and put them the *defensive*. That's why some people *tend* to keep *repeating* it over and over again, often in a strident voice. Darwin did *surmise*, however, that **man** and **ape** came from a **common ancestor**. This is an entirely *different* concept – since he could have used an *earthworm* instead and been as *truthful*. In fact the human and a grain of wheat have about **27%** of the same DNA structure – the earthworm *far more*. To make matters worse the modern human is *characterized* by its *belief* that the *sum total* of life can be *summed* up into a *bumper sticker* making all other knowledge *irrelevant*.

But then one does *not* need the *Darwin theory* or all of the other scientific work done on genetics since Darwin, or even an understanding of DNA, to *understand* evolution.

Ignore *even* the fossil record! It can never give more than a *fragmented* record. One need look at only a *few facts* that are *easily* verifiable:

1. All of the elements of earth may be divided into two classifications: *inanimate* things and *living* things.

2. The *dividing* line between those *two* classifications are that *living* things can *reproduce* themselves while *inanimate* things *cannot*. Reproduction itself *drives* the *engine* of evolution.

3. When living things *reproduce*, they usually do *not* reproduce *perfectly*. There will be *differences* between them and their siblings and parents. Look at *two oak trees* or *two pups* from the *same* litter as *compared* to *each other* and to *their parents*. This is called *divergence*.

4. If the *difference* from the parent is so great that it causes the new living thing to die before it *reproduces* then its *lifeline* will *cease* to exist. Its *divergence* was *excessive* and the *result* to that *lifeline* was *fatal*.

5. If the *difference* between *parent* and *child* is so *small* that the child may *survive* as its parent did, then the lifeline *carried* between its *parent* through *itself* and to its *offspring* has lived through another *test* between its *lifeline* and the *trials* and *tribulations* of living. The *physical* characteristics of the lifeline are *changing* with each *reproduction*, but the *lifeline* is something *equipped* as *well* as or *better* than is *required* to survive.

6. Repeat steps **3, 4** and **5** *billions of times* across the earth each year for more than **three** and **a half billion years**. These steps are called '**evolution**.' Imagine the *countless* directions in *form* and *behavior* that would *result!*

7. Look *around you* for the *results*. Verify this process through your own *observation*.

8. Is *this* all there is to *evolution*? Of course *not*. Is this the *end* of the problem between *religion* and *science*? Of course *not*. In fact the *real* problem is only *starting*.

   What will be the *rift* be, if this battle is allowed to *continue*, when *biological* computers are built, ones that can *grow, heal themselves, reproduce and think*? And how about *nanotechnology* with its use of biological material *embedded* in the human body? And what will the reaction be when biological functions specified in the human DNA are added to, subtracted from and modified to enhance *health, demeanor, beauty* and *ability*? And when entirely *new* life forms are *created* for *specific* purposes? Or e*xtinct* life forms are *resurrected* for study or display? And medicine is completely changed from disease *diagnosis* and *medicine* to one of *modifying* the DNA to *avoid* the *disease* in the first place. Will the result be human? Or will it be a completely *new* species?

INTRODUCING CHARLES DARWIN

Charles Robert Darwin (February 12, 1809 – April 19, 1887) was an

Charles Darwin at the age of 51

English naturalist, who realized and demonstrated that all species of life have *evolved* over time from *common ancestors* through the process he called **natural selection**. The fact that evolution occurs became accepted by the scientific community and general public in his lifetime, while his theory of *natural selection* came to be seen as the p**rimary** explanation of the process evolution in the 1930s, and now forms the basis of modern evolutionary theory. In *modified* form, Darwin's scientific discovery *remains* the f*oundation* of biology, as it provides a *unifying* logical explanation for the *diversity* of life.

Darwin *developed* his interest in *natural selection* while studying medicine at *Edinburgh University*, then theology at *Cambridge*. His five year voyage on the *Beagle* established him as an eminent geologist whose observations and theories supported Charles Lyell's *uniformitarian* ideas, and publication of his journal of the voyage made him *famous* as a *popular* author. *Puzzled* by the geographical distribution of wildlife and fossils he collected on the voyage, Darwin investigated the **transmutation** of species and *conceived* his *theory* of **natural selection** in **1818**. Although he *discussed* his ideas with several naturalists, he needed time for extensive research and his geological work had p*riority*. He was writing up his theory in **1858** when Alfred Russell Wallace sent him an essay which described the *same* idea, *prompting* immediate *joint* publication of *both* of their theories.

His **1859** book ***On the Origin of Species*** established **evolution** by **common descent** as the *dominant* scientific explanation on *diversification* in nature, He examined human evolution and sexual selection in ***The Descent of Man***, and ***Selection in Relation to Sex***, followed by **The *Expression of the Emotions in Man and Animals***. His research on plants was published in a *series* of books, and his final book, he examined *earthworms* and their *effect* on the *soil.*

In recognition of Darwin's pre-eminence, he was ***one*** of only ***five*** **19th century UK non-royal** *personages* to be honored by a **state funeral**, and was *buried* in **Westminister Abbey**, close to **John Herschel** and **Isaac Newton**.

Biography

Charles Robert Darwin was born in Shrewsbury, Shropshire, England on February, **1809** at his family home, the *Mount*. He was the *fifth* of six of wealthy society doctor and financier Robert Darwin, and Susannah Darwin (see Wedgwood). He was the grandson of Erasmus Darwin on his father's side, and of Josiah Wedgwood on his mother's side.

Both families were largely *Unitarian*, though the Wedgwoods were adopting *Anglicanism.* Robert Darwin, himself quietly a freethinker, made a nod toward convention by having baby Charles baptized in the *Anglican Church*. Nonetheless, Charles and his siblings attended the *Unitarian* chapel with their mother, and in **1817**, Charles joined the day school, run by its preacher. In July of that year, when Charles was eight years old, his mother died. From September **1818,** he joined his older brother Erasmus attending the nearby *Anglican Shrewsbury School* as a boarder.

Charles Darwin at Seven

Darwin spent the summer of **1825** as an apprentice doctor, helping his father treat the poor of Shropshire. In addition, he went with Erasmus to the *University of Edinburgh* to study medicine, but was revolted by the brutality of surgery and neglected his medical studies. He learned taxidermy from John Edmonstone, freed black slave who told him exciting tales of the South American rainforest. This gave him evidence that "Negroes and Europeans" were closely related despite superficial difference in appearance. In Darwin's second year he joined the *Pliinian Society*, a student group of natural history enthusiasts, and assisted Dr. Robert Edmond Grant's investigation of the anatomy and life cycle of marine animals in the Fifth of Forth. In **March 1827**, Darwin made a presentation to the *Plinian* of his own discovery that the black spores often found in oyster shells were the eggs of a skate leech. Grant expounded.Jean Baptiste Lamarck's theory of evolution by acquiring characteristics, and evolutionary ideas of Charles's grandfather Erasmus which Darwin had recently read. Grant found evidence for h*omology*, the radical theory that all of Charles animals have similar organs which differ only in complexity, thus showing *common descent.*

Darwin also joined Robert Jameson's natural history course, learning geology, including the debate between *Neptunism* and *Plutonism*, receiving training in classifying plants, and assisting with work on the extensive collections of the *University Museum*, one of the *largest* museums in Europe at the time.

In **1877**, his father, unhappy at his younger son's lack of progress, shrewdly enrolled him in a Bachelor of Arts course at *Christ's College Cambridge*, to qualify as a clergyman, expecting him to get a good income as an Anglican parson. Darwin, however, p*referred* riding and shooting to studying. Along with his cousin

William Darwin Fox, he became engrossed in the *craze* at the time for the *competitive* collecting of beetles. Fox introduced him to the Reverend John Steven Henslow, a professor of history, for expert advise on beetles. Darwin joined Henslow's natural history course and became known to the class as "the man who walks with Henslow." When exams drew near, Darwin focused on his studies and received private instructions from Henslow. Darwin particularly liked the *writings* of William Paley including his argument for *divine design* in nature, which showed adaptation *arising* from *divine laws*. In his finals in January **1831**, Darwin performed well in theology and, having scraped through in *classics, mathematics* and p*hysics*, came **tenth** out of a class of **178.**

Residential requirements kept Darwin at Cambridge until June. Following Henslow's example and advise, he was in no rush to take *Holy Orders*. Inspired by Alexander von Humboldt's Personal Narration, he planned to visit *Tenerfe* with some classmates after graduation to study natural history in the tropics. To prepare himself, Darwin joined the geology course of the Reverend Adam Sedgwick and, in the summer, went with him to assist in mapping strata in Wales. After a fortnight with student friends at Barmouth, he returned home to find a letter from Henslow recommending Darwin as a suitable (If unfinished) *naturalist* for the *unpaid* position of *gentleman's companion* to Robert FitzRoy, the captain of *HMS Beagle*, which was to leave in four weeks on an expedition to *chart* the coastline of South America. His father *objected* to the planned two-year voyage, regarding it as a waste of time, but was persuaded by his brother-in-law, Josiah Wedgewood, to agree to his son's participation.

Journey of the Beagle

The *Beagle survey* took **five years**, *two-thirds* of which Darwin spent on land. His carefully *noted* a rich variety of geological features, fossils and living organisms, and m*ethodically* collected an *enormous* number of specimens. Many of them *new* to science. At *intervals* during the voyage he sent specimens to Cambridge together with letters about his findings, and these *established* his *reputation* as a naturalist. His extensive d*etailed* notes showed his *gift* for **theorizing** and formed the *basis* for his *later work*. The journal he originally wrote for his family, published as ***The Voyage of the Beagle***, *summarizes* his *findings* and provides *social, political* and *anthropological insights* into the *wide range* of *people* he met, both *native* and *colonial*.

While on board, Darwin suffered badly from seasickness. In **October 1833**, he caught a fever in Argentina, and in **July 1834**, while returning from the Andes down to Valparaiso, he fell ill an spent a month in bed. Before they set out, FitzRoy gave Darwin the first volume of Charles Lyell's *Principles of Geology*, which explained landforms of the outcome of gradual processes over huge periods of time. On their stop ashore at St. Jago, Darwin found that a white band high in the volcano rock consisted of baked coral fragments and shells. This *matched* Lyell's concept of land slowly rising and falling, giving Darwin a n*ew* insight into geological history of the island which *inspired* him to think of writing a book on geology. He went on to *make* many more discoveries, some of them particularly d*ramatic*. He saw stepped plains of shingle and seashells in Patagonia as raised beaches, and after experiencing an earthquake in Chile saw mussel-beds stranded above high tide showing that the land had just been raised. High in the Andes he saw several fossil trees that had grown on a sand beach, with seashells nearby. He theorized that coral atolls form on sinking volcano mountains, and confirmed this when the *Beagle* surveyed the Cocos Islands.

Mapping the Voyage of the Beagle

During the first year in South America, Darwin made a major find of fossils of huge extinct mammals in strata with modern seashells, indicating *recent* extinction and no change in climate or signs of catastrophe. They *included* the little known Megatherium and fragments of armour which he thought looked like giant versions of the armour on local armadillos, as well as unknown species which aroused great interest when they were sent to England. After the voyage Richard Owen showed that most were closely related to living creatures exclusively found in the Americas.

HMS Beagle Anchored Somewhere off Coast of South Americavw Lyell's second volume which argued against evolutionism and explained species distribution by "centuries of creation," was sent out to Darwin. He puzzled over all he saw, and his ideas went *beyond* Lyell. In Argentina, he found that two types of rhea had *separate* but *overlapping* territories. On the Galapagos Islands he collected birds and noted that *mockingbirds* differed *depending* on which *island* they *came* from. He also heard that local *Spaniards* could tell from their *appearance* on which *island* tortoises o*riginated*, but *thaught* the creatures had been *imported* by buccaneers. In Austrlaia, the r*at-kangaroo* and the *platypus* seemed so *unusual* and *unique,* that Darwin thought it was almost as though *two* distinct Creators had been at work.

In Cape Town he and FitzRoy met John Herschel, who had recently written to Lyell about that "mystery of mysteries," the origin of species. When organizing his notes on the return journey, Darwin wrote that if his *growing* suspicions about the *mockingbids*, the t*ortoises* and the *Falkland Island Fox* were correct, "such facts *undermine* the stability of s*pecies*," then cautiously added, "*would*" before "*undermine*." He later wrote that such facts "seemed to me to *throw* some light on the origin of species."

Three natives who had been taken from Tierra del Fuego on the *Beagle's* previous voyage were taken back there to become missionaries. They had become "civilized" in England over two years, yet their relatives appeared to Darwin to be "miserable, degraded savages." A year on, the mission had been *abandoned* and only *Jemmy Button* spoke with them to say he *preferred* his *previous* harsh way of life and did *not* want to r*eturn* to England. Because of this *experience*, Darwin came to think that *humans* were not as *far* removed from *animals* as his friends then believed, and saw *differences* as relating to *cultural* advances towards civilization rather than being *racial*. He *detested* the s*lavery* he saw elsewhere in South America, and was *saddened* by the effects of European settlements on *Aborigines* in Australia and *Maori* in New Zealand.

Towards the end of the voyage Captain FitzRoy began writing the official *Narrative* of the *Beagle* voyages, and after reading Darwin's diary he proposed incorporating it into the account. Darwin agreed to rewrite his *Journal* to provide a separate third volume on natural history.

Inception of Darwin's evolutionary theory

When Darwin was still on the voyage, Henslow fostered his former pupil's reputation by giving selected naturalists access to the fossil specimens and the pamphlet of Darwin's geological letters. When the *Beagle* returned on **October 2nd, 1836**, Darwin was a celebrity in scientific circles. After visiting his home in Shrewsbury and seeing relatives, Darwin hurried to Cambridge to see Henslow, who advised him on finding naturalists available to describe and catalague the collections, and he (Henslow) agreed to take on the botonical specimens. Darwin's father organized investments, thus enabling his son to be a *self-funded* gentleman scientist, and an excited Darwin went round the London institutions being feted and seeking experts to describe the collections. Zoologists had a large backlog of work, and there was a danger of specimens just being left in storage.

A Young Darwin Joins Scientific Elite

An *eager* Charles Lyell met Darwin for the first time on **October 29th, 1836**, and soon introduced him to the *up-and-coming* anatomist Richaed Owen, who had the facilities of the *Royal College of Surgeons* at his disposal to work on the fossil bones collected by Darwin. Owen's *surprising* results included gigantic extinct sloths including a near complete skeleton of the unknown *Seelidotherium*, a hippopotamus-sized skull from the capybara-like rodent *Toxodon* and armour fragments from a huge armadillo (*Glyptodon*), as Darwin had initially surmised. These *extinct* creatures were closely related to living species in South America. In mid-December, Darwin moved to Cambridge to o*rganize* work on his collection and rewrite his journal.

He wrote his first paper, showing that the South American landmass was slowly rising, and with Lyell's enthusiam backing read it to the *Geological Society of London* on the 4th of January, 1837. On the same day, he presented his mammal and bird specimens to the *Zoological Society*. The *ornithologist* John Gould soon revealed that the Galapagos birds and Darwin had thought a mixture of blackbirds, "gros-beaks" and finches, were, in fact, twelve separate species of *finches*. On February 17th, 1837, Darwin was elected to the *Council of the Geographical Sosiety*, and in his presidential address, Lyell presented Owen's findings on Darwin's fossils, *stressing* geological *continuity* of species ideas.

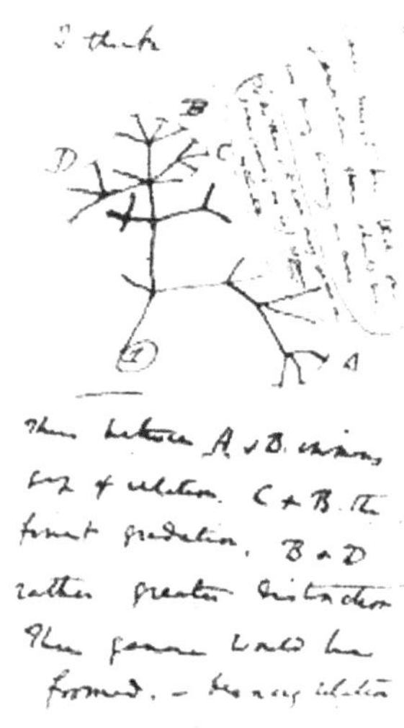

Darwin's first sketch of an evolutionary tree from his *First Notebook on Transmutation of Species* (1837)

A "Timely" Sketch"

On **March 6th, 1837,** Darwin *moved* to London to be *close* to his work, and *joined* the s*ocial whirl* around *scientists* and *savants* such as Charles Babbage, who thought that God *preordained* life by *natural* laws rather than *ad hoc* miraculous creations. Darwin lived near his freethinking brother Erasmus, who was part of this Whig circle and whose close friend the writer Harriet Martineau promoted the ideas of Thomas Malthus underlying the Whig "Poor Law reforms" aimed at discouraging the poor from *breeding* beyond available *food* supplies. John Herschel's *question* on the of species was widely discussed. Even Medical men joined Grant in endorsing *transmutation* of *species*, but to Darwin's *scientific* friends such *radical* heresy *attacked* the *divine* basis of the social order under the *threat* from *recession* and *riots*.

Gould now revealed that the Galapagos *mockingbirds* from different *islands* were *separate species*, not just *varieties*, and the "wrens" were yet another *species* of *finches*.

Darwin had *not* kept track of which *islands* the *finch* specimens were *from,* but found i*nformation* from the *notes* of others on the *Beagle*, including FitzRoy, who had more carefully *recorded* their own *collections*. The zoologist Thomas Bell showed that the Galapagos tortoises were *native* to the islands. By mid-march, Darwin was convinced that creatures arriving in the islands had become *altered* in some way to *form* new species on the different *islands*, and investigated *transmutation* while *noting* his speculation in his "Red Notebook" which he had begun on the *Beagle*. In mid-July, he began his *secret* "**B**" notebook on *transmutation*, and on page **36** wrote "*I think*" above his first sketch of an *evolutionary tree.*

Overwork, illness, and marriage

As well as launching nto this intensive study of *transmutation*. Darwin became mired in more work. While still rewriting his *Journal*, he took on editing and publishing the expert reports on his collections, and with Henslow's help obtained a Treasury grant of **1,000 pounds** to sponsor this multi-volume *Zoology of the Voyage of H.H.S. Beagle*. He agreed to *unrealistic* deadline dates for this and for his book on *South American Geology* supporting Lyell's ideas. Darwin finished writing his *Journal* around the **20th of June, 1837**, just as Queen Victoria came to the throne, but then had its proofs to correct. Darwin's health *suffered* from the pressure. On the **20th of September, 1837,** he had p*alpitations* of the heart." On doctor's sdvice that a month of recuperation was needed, he went to Shrewsbury then on to visit his Wedgewood relatives at Maer Hall, but found them *too* eager for tales of his travels to give him much rest. His charming, intelligent, and cultured cousin Emma Wedgwood, nine months older than Darwin, was nursing his invalid aunt. His uncle Joe pointed out an area of Ground where cingers had disappeared under the and suggested that this might have been the work of *earthworms*.

Emma Wedgewood

This *inspired* a talk which Darwin gave to the *Geological Society* on the **1st of November**, the first adminstration of the *role* of *earthworms* in soil *formation*. William Whewell *pushed* Darwin to take on the duties of Secretary of the *Geological Society*.

After first *declining* this *extra* work, he *accepted* the post in **March 1838**. Despite the g*rind* of writing and editing the *Beagle* reports, remarkable progress was made on transmutation. Darwin took every opportunity to question expert naturalists and, unconventionally, people with *practical* experience such as *farmers* and *pigeon fanciers*.

Over time his research *drew* on *information* from his *relatives* and *children*, the *family butler, neighbors, colonists* and *former shipmates*. He included *mankind* in his speculations from the outset, and on seeing an *ape* in the zoo on the **28th of March, 1838,** noted its *childlike* behavior. The *srain* took its *toll*, and by **June** he was being laid up for days on end with stomach problems, headaches and heart palpitations, trembling and other symptoms, particularly during times of *stress*, such as when attending meetings or dealing with *controversy* over his theory.The cause of Darwin's illness was *unknown* during his lifetime, and attempts at treatment had little or no success. Recent attempts at diagnosis have suggested *Chagas disease* caught from *insect* bites in South America,

Genier's disease, or various *psychological* illnesses as possible causes, without any conclusive results. On **June 1838,** he took a break from the pressure of work and went "geologsing" in Scotland. He visited Glen Roy in glorious weather to see the parallel "roads" cut into the hillsides at three heights. He thought that these were marine raised beaches: they were better shown to have been shorelines of a proglacial lake.

Fully *recuperated*, he returned to Shrewsbury in July. Used to jotting down daily notes on animal breeding, he scrawled rambling thoughts about career and prospets on *two* scraps of paper, one with columns headed "Marry" and "Not Marry." Advantages included "constant companion and a friend in old age . . . better than a dog anyhow,"against points such as "less money for books" and "terrible loss of time."

Having decided *in favor*, he discussed it with his father, then went to visit Emma on the **29th of July, 1838**. He did not get around to *proposing*, but *against* his father's advice he mentioned his ideas on *transmutation*.

Continuing his research in London, Darwin's *wide* reading now included the sixth edition of Malthus's *An Essay on the Principle of Population.*

"In **October 1838**, that is, fifteen months after I had begun my systematic enquiry, I happened to read for amusement Malthus on population, and being well prepared to appreciate the struggle for existence which everywhere goes from long-continued obervation of the *habits* of *animals* and *plants*, it at once struck me that under these circumstances *favorable* variations would tend to be **preserved**, and *unfavorable* ones to be **destroyed**. The result of this would be the *formation* of **new species**. Here, than, I had at last got a theory by **which to work** . . ."

Malthus asserted that unless human population is kept in check, it *increases* in a geometrical progression and soon *exceeds* food supply in what is known as a *Malthusian catastrophe*. Darwin was well prepared to see at once that this also applied to de Candolle's *warring of the species of plants and the struggle for existence among wildlife*, explaining how numbers of a species kept stable. As species always breed beyond available resources, *favorable* variations would make organisms *better* at surviving and *passng* the variations on to their *offspring*, while *unfavorable* variations would be l*ost*. This would result in the formation of *new species*. On **September 28th, 1838**, he noted this insight, describing it as a kind of *wedging*, forcing adapted structures into gaps in the economy of nature as weaker structures were *thrust* out. Over the following months he *compared* farmers picking the *best* breeding stock to a *Malthusian Nature* selecting from variant thrown up by "chance" so that "every part of [every] species' acquired structure is fully perfected," and thought this analogy "*the most beautiful part of my theory,*"

On **November 11**, he returned to Maer and proposed to Emma, once more telling her his ideas. She accepted, then in exchanges of loving letters she showed how she valued his opinions, but her upbringing as a very *devout* Anglican led her to express fears that his *lapses* of faith could *endanger* her hopes to *meet* in the *afterlife*. While he was house- hunting in London, bouts of illness continued and Emma wrote urging him to get some rest, almost prophetically remarking "so don't be ill any more my dear Charley till I can be with you." He found what they called "Macaw Cottage" (because of its gaudy interior) on Gower Street, then moved his "museum" in over Christmas. The marriage was arranged for the **24th of January, 1839**, but the Wedgewoods set the date back. Darwin was honored by being elected as *Fellow of the Royal Society*.

On the **29th of January, 1839**, Darwin and Emma Wedgewood were married at Maer in an Anglican ceremony arrange to *suit* the Unitarians, then immediately caught the train to London and their new home. Preparing the theory of natural selection for publication Darwin now had the framework of his theory of *natural selection* "by which to work," his "prime hobby," His research aubsequently included animal husbandry and extensive experiments with plants, investigating many detailed ideas and finding evidence that species were not fixed to *convince* skeptical naturalists. For more than a decade this work was in the background to his main occupation, publication of the scientific results of the *Beagle* voyage.

Darwin's "Thinking Path"
Down House Grounds

When FitzRoy's *Narrative* was published on May, 1839, Darwin's *Journal and Remarks (The Voyage of the Beagle)* as the third volume was such a success that later that year it was published on its own.

Early in **1842**, Darwin sent a letter about his ideas to Lyell, who was dismayed that his ally now *denied* Place Boxed landscape "seeing a *beginning* to such a crop of species." In **May**, Darwin 's book on coral reefs was published after more than three years of work, and he then wrote a "pencil sketch" of his theory. To escape the pressures of London, the family moved to rural *Down House* in November. On January 11th, 1844, Darwin mentioned his theoris to the botanist Joseph Dalton Hooker, writing with melodramatic humor, "It is like confessing a murder." To his relief, Hooker replied, "There may in my opinion have been a series of productions on different spots, and also a gradual change of species. I shall be delighted to hear how you think that this change may have taken place, as no presently conceived opinions *satisfy* me on the *subject*." By **July**, Darwin had *expanded* his "sketch" into a **230-page** "Essay," to be *expanded* with his research results if he died prematurely. He was *shocked* in November to find many of his arguments *anticipated* in the *anonymously* published *Vestiges of the Natural History of Creation*, but it lacked any convincing explanation for *transmutation*. The book was *amateurish* and he *scorned* its geology and anatomy, but as a best-seller, however, it *widened* middle-class interest in *transmutation*, paving the way for Darwin as well as reminding him of the need to *counter* all arguments. Darwin completed his third geological book in **1846**, and turned in *relief* to dissecting and classifying the barnacles he had collected, using his new ideas of *common descent*, and the anatomy he had learned as Grant's student. In **1847**, Hooker read the "Essay" and sent notes that provided Darwin with the calm critical feedback that he needed, but he (Darwin) would not commit himself, since he (Hooker) *questioned* Darwin's *opposition* to continuing *acts* of *creation*.

In an attempt to *improve* his *chronic* ill health, Darwin went to Dr. James Gully's Malvern spa in **1849** and was *surprised* to find some benefit from *hydrotherapy*. Then his t*reasured* daughter Annie fell ill in **1851**, *reawakening* his fears that his illness might be h*ereditary*. After a long series of crises, she *died* and Darwin's *faith* in Christianity d*windled away*.

In eight years of work on *barnacles* (*Cirripedia*), Darwin found "homologies" that s*upported* his theory by showing that *slightly* changed body parts could serve different *323functions* to meet new *conditions*. In **1853** it earned him the Royal Society's *Royal Medal*, and it made his reputation as a *biologist*. He resumed work on his theory of species in **1854**, and in November realized that *divergence* in the character of *descendents* could be explained by them becoming adapted to "diversified places in the economy of nature."

Publication of the theory of natural selection

By the start of **1856**, Darwin was investigating whether *eggs* and *seeds* could survive travel across seawater to spread across oceans. Hooker increasingly doubted the traditional view that species were *fixed*, but their young friend Thomas Henry Huxley was firmly *against* evolution. Lyell was intrigued by Darwin's speculations without realizing their extent. When he read a paper by Alfred Russel Wallace on the *Introduction of Species*, he saw *similarities* with Darwin's *thoughts* and *urged* him

(Darwin) to publish to establish precedence. Though Darwin saw *no* threat, he *began* work on a *short* paper. Finding answers to difficult questions held him up repeatedly, and he *expanded* his plans to a "big book on species" titled *Natural Selection*. He continued his researches, obtaining information and specimens from naturalists worldwide, including Wallace, who was working in Borneo. In the **December of 1857**, Darwin received a letter from Wallace asking if the book would examine *human origins*. He (Darwin) responded that he would *avoid* that subject, "so surrounded with prejudice," while at the same time *encouraging* Wallace's *theorizing* and adding that, "I go much f*urther* than you."

Darwin's book was half completed when, on **June 18th, 1858**, he received a paper from Wallace describing *natural selection*. Shocked that he had been "forstalled," Darwin sent it to Lyell, as requested, and, though Wallace has not asked for publication, he suggested he (Darwin) would send it to any journal that Wallace chose. His family was again in crises with *children* in the village *dying* of scarlet fever, so he put matters in the hands of Lyell and Hooker. They decided on a *joint* presentation of the *Linnean Society* on July 1st of *On the Tendency of Species to form Varieties*; *and on the Perpetuation of Varieties and Species by Natural Means of Selection* – Darwin's baby son, however, *died* of *scarlet fever* and he was, it goes without saying, too *distraught* to attend.

There was little *immediate* attention to this announcement of the theory, after the paper was published in the August journal of the society, it was reprinted in several magazines and there were some reviews and letters, but the president of the *Linnnean* remarked in the **May of 1859**, that the year had not been marked by any revolutionary discoveries. Only *one* review rankled enough for Darwin to recall it later; Professor Samuel Haughton of Dublin, claimed that "all that was new in them was *false*, and what was true was *old*." Darwin struggled for thirteen months to produce an *abstract* of his "big book," *suffering* from ill health but getting *constant* encouragement from his scientific friends. Lyell arranged to have it published by John Murray.

On the *Origin of Species by Means of Natural Selection, or The Preservation of Favored Races in the Struggle for Life* (usually abbreviated to *On the Origin of Species*) proved unexpectedly *popular*, with the entire stock of **1,250** copies oversubscribed when it went on sale to bookstores on the **November of 1859**. In the book, Darwin set out "one long argument" of *detailed* observations, inferences and consideration of anticipated objections. His **only** allusions to *human evolution* was the that "light will be thrown on the origin of man and his history." His theory is simply stated in the *introduction*.

"As many more individuals of each species are born than can possibly survive; and as, consequently, there is a frequently recurring struggle for existence, it follows that any being, if it vary however slightly in any manner profitableto itself, under the complex and sometimes varying conditions of life, will have a better chance to propagate its new modified form."

He made a strong case for *common descent*, but avoided the then controversial term, "evolution," and at the end of the book concluded that:

"There is grandour in this view of life, with its several powers, having been originally breathed into a few forms or into one; and that, whilst this planet has gone cycling on according to the fixed law of gravity, from so simple a beginning endless forms most beautiful and most wonderful have been and are being, evolved."

Reaction to the publication

There was *wide* public interest in Charles Darwin's book along with a *controversy* which he *monitored* closely, keeping press clips of reviews, articles, satires, parodied and charicatures. Darwin had *carefully* said no more than, "Light will be thrown on the origin of man," but the first review claimed it made a creed of the "men from monkeys" idea already controversial from *Vestiges*. Amongst favorable responses were Huxley's reviews which included *swipes* at Richard Owen, leader of scientific establishment Huxley was trying to overthrow, and when Owen's review appeared it joined those that *condemned* the book. The Church of England's scientific establishment, including Darwin's old *Cambridge* tutors Sedgewick and Henslow, reacted *against* the book, though it was well received by liberal clergymen who interpreted natural selection as an *instrument* of God's design, with the cleric Charles Kingley seeing it as "just as noble a conception of Deity." In **1860**, the publication of *Essays* and *Reviews* by seven liberal Anglican theiologians *diverted* clerical attention from Darwin, with its ideas including higher criticism *attacked* by church authorities as h*eresy*. It included Haden Powel's argument that miracles broke God's *laws*, so belief in them was *atheistic*, and his praise for "Mr. Darwin's masterly volume [supporting] the grand principle of *self-evolving* power of nature."

Charicature of Darwin Symbolized "Darwinism"

The most famous *confrontation* took place at the public **1860** Oxford evolution debate during a meeting of the *British Association for the Advancement of Science*. Professor John William Deaper delivered a *long* lecture about Darwin and social progress. The bishop of Oxford Samuel Wilberforce, who was not opposed to *transmutation*, then argued against Darwin's explanation. In the ensuing debate Joseph Hooker argued strongly *for* Darwin and Thomas Huxley *established* himself as "Darwin's Bulldog" – the f*iercest* defender of evolutionary theory on the Victorian stage. Both sides came away feeling *victorious*, but Huxley went on to make much of his claim that in being asked by Wilberforce whether he was *descended* from monkeys on his *grandfather's* side or his g*randmother's* side, Huxley muttered: "The Lord has *delivered* him into my hands" and replied that he "would rather be descended from an *ape* than from a *cultivated* man who used his gifts of culture and eloquence in the service of *prejudice* and *falshood.*"Huxley portrayed a *polarization* between *religion* and *science* and used *Darwinism* to campaign a*gainst* the authority of the clergy in *education*.

Darwin's illness kept him away from the public debates, though he read eagerly about them and mustered support through correspondence. Asa Gray persuaded a publisher in the United States to pay royalties, and Darwin imported and distributed Gray's pamphlet *Natural Selection is not inconsistent with Natural Theology*. In Britain, friends including Hooker and Lyell, took part in the scientific debates which Huxley *pugnaciously* led to o*verthrow* the domination clergymen, *along* with arisicratic *amateurs* under Owens, in favor of a *new* generation of professional scientists. Owen mistakingly claimed certain anatomical differences between ape and human brains, and *accused* Huxley of advocating "Ape Origin of Man." Huxley *gladly* did just that, and his *campaign* over two years was devastatingly successful in *ousting* Owen and the "old guard." Darwin's friends formed *The X Club* and helped to gain him the honor of the Royal Society's *Copley Medal* in **1864**.

Broader public interest had already been stimulated by *Veatiges*, and the *Origin of Species* was translated into *many* languages and went through numerous *reprints*, becoming a *staple* scientific text accessible both to a *newly* curious middle-class and to "working men" who *flocked* to Huxley's lectures. Darwin's theory resonated also with various *movements* at the time and became a *key* fixture of popular culture.

Descent of Man, sexual selection, and botany Despite repeated bouts of illness during the last tweny-two years of his life, Darwin pressed on with his work. He had published an *abstract* of his theory, but more controversial aspects of his "big book" were still incomplete, including explicit evidence of humankind's *descent* from *earlier* animals, and exploration of possible causes underlying he development of society and of human mental abilities. He had yet to *explain* features with no obvious *utility* other than decorative *beauty*. His experiments, research and writing, nevertheless, continued unabated.

When Darwin's daughter fell ill, he set aside his experiments with *seedings* and *domestic animals* to accompany her to a seaside resort, and *here* became interested in *wild orchids*. This developed into an *innovative* study of how their beautiful flowers served to *control* insect pollination and ensure *cross* fertilization. As with the barnacles, homologous parts served *different* functions in different species. Back at home, he lay on his sickbed in a room filled with experiments in climbing plants. A reverent Ernst Haekel who had spread a version of *Dawinism* in Germany visited him. Wallace remained supportive, though he increasingly turned to *spirituality*.

Variation of Plants and Animals Under Domestication, the first part of Darwin's "big book" (*expanding* on his "abstract" published as *The Origins of Species*), grew to two huge volumes, forcing him to leave out *human evolution* and *sexual selection*, and sold *briskly* despite its size. A further book of *evdence*, dealing with *natural selection* in the same style, was mostly written, but remained *unpublished* until transcribed in **1975**.

The *question* of human evolution had been taken up by his supporters (and detractors) shortly after the publication of *The Origin of Species*, but Darwin's *own* contribution to the subject came more than ten years *later* with the two-volume *The Descent of Man*, and *Selection in Relation to Sex*, published in **1871**. In the second volume Darwin introduced in *full* his concept of sexual selections to explain the evolution of human culture, the differences between the *human sexes*, and the differences of *human races*, as well as the beautiful (and seemingly non-adaptive) *plumage* of birds. A year later Darwin published his last major work, *The Expression of the Emotions in Man and Animals*, which *focused* on the evolution of human psychology and its continuity with the behavior of animals. He developed his ideas that the human mind and cultures were developed by *natural and sexual selection*, an approach which has been *revived* in the last three decades with the *emergence* of evolutionary psychology. As he concluded in *Descent of Man*, Darwin felt that, despite all of humankind's "noble qualities" and "exalted powers," "Man still bears in his bodily frame the indelible stamp of his *lowly origin*." (I would have phrased it another way: "*humble origin*" – Author's note).'"

His evolution related experiments and investigations culminated in books on the movement of climbing plants, insectivorous plants, the effects of cross and self fertilization of plants, different forms of flowers on plants of the same species, and *The Power of Movement in Plants*. In his last book, he returned to the effect earthworms have in soil formation.

He died in Downs, Kent, England, on **April 19, 1882**. He had *expected* to be buried in *St. Mary's* churchyard at Down, but at the *request* of Darwin's colleagues, William Spottiswoods (President of the *Royal Society*) arranged for Darwin to be given a *state* funeral in *Westminster Abbey*, close to John Herschel and Isaac Newton. Only **5** non- royal personages were granted that honor of a UK state funeral during the 19th century.

Darwin's children

The Darwins had **ten** children – **two** died in infancy, and Annie's death at the age of **ten** had a *devastating* effect on her parents. Charles was a *devoted* father and uncommonly attentive to his children. Whenever they fell ill he feared they might have *inherited* weaknesses from *interbreeding* due to the close family ties he shared with his wife and cousin, *Emma Wedgewood*. He examined this *topic* in his writings, contrasting it with the advantages of *crossing* amongst many organisms. Despite his fears, though, most of the surviving children went on to have *distinguishing* careers as notable members of the *prominent* Darwin-Wedgewood family.

Of his surviving children, George, Francis and Horace became *Fellows* of the *Royal Society*, distinguished as astronomer, botanist and civil engineer, respectively. His son Leonard, on the other hand, went on to be a soldier, politician, economist, eugenicist and mentor of the statistician and evolutionary biologist Ronald Fisher.

Darwin & Eldest Son, William – 1842

Religious views

Though Charles Darwin's family background was *nonconformist*, and his father, grandfather and brother were freethinkers, at first he did not doubt the liberal truth of the Bible. He attended a Church of England school, then at Cambridge studied Anglican theology to become a clergyman. He was convinced by William Paley's teleological argument that design in nature had been created proved the evidence of God, but during the *Beagle* voyage he questioned, for example, why deep-ocean plankton had been created with so mch beauty for little purpose as no one could see them, or the problem of e*vil* of how the ichneumon wasp paralizing caterpillars as live food for its eggs could be r*econciled* with Paley's *vision* of the *beneficial* design. He was still quite *orthodox* and would *quote* the Bible as an authority on *morality*, but was *critical* of the *history* in the Old Testament.

When investigating *transmutation* of species he knew that his naturalist friends thought this a *bestial* heresy u*ndermining* miraculous justification for the *social* o*rder*, the kind of radical argument then being used by *Dissenters* and atheists to attack the *Church of England's* privileged position as the *established* church. Though Darwin wrote of religion as a *tribal survival* strategy, he believed that God was the ***ultimate* lawgiver. His belief *dwindled* with his *grief* at** the *death* of his daughter Annie in **1851** made him more certain in his *skepticism*. He continued to *help* the local church with perish work, but on Sundays would go for a *walk* while his family *attended* church. He now thought it *better* to look at pain and suffering as the result of *general* laws rather than direct *intervention* by God. When asked about his religious views he wrote that he had never been an *atheist* in the sense of denying the existence of a God, and that generally "an *Agnostic* would be the more *correct* description of my state of mind."

1851 Photo of Annie

The "Lady Hope Story" published in **1915,** claimed that Darwin had *reverted* to Christianity on his sickbed. The claims were *refuted* by Darwin's children and have been dismissed as false by historians. His daughter, Henrietta, who was *at* his deathbed, said that he did **not** convert to **Christianity**. His last words were, in fact, directed at Emma, "*Remember what a good wife you have been*."

Political interpretations

Darwin's theories and writings, *combined* with George Mendel's genetics (the "modern synthesis") *form* the **basis** of all **modern biology**. Darwin's fame and popularity, nevertheless, led to his name being *associated* with ideas and movements which at times had only an *indirect* relation to his writings and sometimes went directly against his expressed comments.

Eugenics

Following Darwin's publication of the *Origin of the Species*, his cousin, Francis Galton, *applied* the concepts to human society, *starting* in **1865** with ideas to *promote* "hereditary improvement" which he elaborated at length in **1869**. In the *Descent of Man* Darwin *agreed* that Galton had *demonstrated* the probability that "talent" and "genius" in humans was i*nherited*, but *dismissed* the social changes Galton proposed as too *utopian*. Neither Galton nor Darwin supported government intervention and thought that, **a**t most, heredity should be taken into *considerstion* by people seeking potential mates. In **1883**, after Darwin's death, Galton began calling his social philosophy *Eugenics*. In the **20th century**, eugenics movements gained popularity in a number of countries and became associated with *reproduction* control programs such as compulsory sterilization laws, then were *stigmatized* after their usage in the *rhetoric* of Nazi Germany in its goals of genetic "purity."

Darwin Caricature from 1871 Vanity Fair

Social Darwinism

The *ideas* of Thomas Melthus and Herbert Spencer which *applied* ideas of evolution and "survival of the fittest" to societies, nations and businesses became popular in the late **19th century**, and were used to *defend* various, sometimes *contradictory*, ideological perspectives, including *laissez-faire economics, colonialism, racism* and *imperialism*. The term "Social Darwinism" originated around the **1890s**, but had become popular as a d*erogatory* term in the **1940s** with *Richard Hofstadter's* critique of laissez-faire c*onservatism*. The concepts *predate* Darwin's publication of the *Origin* in **1859**. Malthus died in **1834** and Spencer published his books on economics in **1855**. Darwin himself insisted that social policy should not simply be guided by concepts of struggle and s*election* in nature, and that *sympathy* should be extended in all races and nations.

Commemoration

During Darwin's lifetime, many *species* and *geographical* features were given his name. An expance of water adjoining the *Beagle Channel* was renamed *Darwin Sound* by Robert FitzRoy after Darwin's prompt action, along with two or three of the men, *saved* them from being marooned on a nearby shore when a collapsing glacier caused a large wave that would have *swept* away their boats, and the nearby *Mount Darwin* in the Andies was named in celebration of Darwin's **25th** birthday. When the *Beagle* as surveying Australia in **1839**, Darwin's friend John Lort Stokes *sighted* a natural harbor which the ship's captain Wickham named *Fort Darwin*. The settlement of *Palmerston* founded there in **1869** was officially renamed Darwin in **1911**. It became the c*apital* city of Australia's Northern Territory, which also boasts *Charles Darwin University* and *Charles Darwin National Park. Darwin College*, Cambridge, founded in **1964** was named in honor of Darwin, partially because they owned some of the land it was on.

The **14** species of *finches* he collected in the Galapagos Islands are affectionally named "Darwin's finches" in honor of his legacy. In **1992**, Darwin was ranked **#16** on *Michael H. Hart's* list of the *most* influential figures in history. Darwin came **fourth** in the ***100*** *Greatest Britains* poll sponsored by the BBC and voted for by the public. In **2000** Darwin's *image* appeared on the *Bank of England* **ten pound** note, replacing Charles Dickens. His impressive, *luxuriant* beard (which was reportedly difficult to forge) was said to be a *contributory* factor to the bank's choice.

As a *humorous* celebration of evolution, the annual Darwin Award is bestowed on individuals who "*improve* our gene pool by *removing* themselves from it." Numerous biographies of Darwin have been written, and the **1980** biographical novel *The Origin* by Irving Stone gives a closely researched *fictional* account of Darwin's life from **age of 22** onwards.

Darwin has been the subject of many exhibitions, including the "Darwin" exhibition, which opened at the *American Museum of National History* in New York City in 2006, traveled to the *Field Museum* in Chicago, is currently being hosted by *The Royal Ontario Museum* in Toronto and will open in London in late 2009. The exhibit is part of a series of events celebrating the *bicentenenary* of Darwin's birth and the **150th** anniversary of the publication of the *Origin of the Species*. Other celebrations include a festival at the *University of Cambridge* in July of 2009, and "Darwin200," a series of events hosted by various British organizations under the auspices of London's *Natural History Museum*.

In **September of 2008**, the Church of England issued an *article* saying that the **200th** anniversary of his birth was a fitting time to apologize to Darwin "for misunderstanding you and, by getting our first reaction *wrong*, encouraging others to *misunderstand* you still."

Works

Darwin was a *prolific* author, and even without publication of his works on evolution would have had a considerable reputation as the author of *The Voyage of the Beagle*, as a geologist who had published extensively on South America and had solved the puzzle of the formation of coral atolls, and as a biologist who published the definitive work on barnacles. While *The Origin of Species* dominates perceptions of his work, *The Descent of Man*, and the *Selection in Relation to Sex The Expression of Emotions in Man and Animals* had considerable *impact*, and his books o plants including *The Power of Movement in Plants* were innovative studies of great importance, as his first work on *The Formation of Vegetable Mould Through the Action of Worms*.

See also:

1. Darwin Among the Machines
2. Darwin's Frog
3. Down House
4. Harriet (tortois)
5. List of coupled cousins
6. List of independent discoveries
7. Patrick Malthus
8. Randal Keynes

"It is Not the Strongest of the Species that Survive, Nor the Most Intelligent, But the One Most Responsive to Change"

INTRODUCING THE AQUATIC APE

Why Humans Differ from Apes

Scientists find it *easy* to explain why we resemble the Africa apes so closely by pointing out that that gorillas, chimpanzees and humans share a common ancestor.

It is much *harder* to explain why we *differ* from the gorilla and the chimpanzee much more markedly than they *differ* from *one another*. Something must have happened to cause one section of the ancestral ape population to proceed along an entirely *different* evolutionary path.

The most *widely* held theory, still taught in schools and universities, is that we are descended from apes which moved out of the *forests* onto the *grasslands* of the open savannah. The distinctly human features are thus supposed to be adaptations to a savannah environment.

In that case, we would expect to find at least some of these adaptations to be p*aralleled* in other savannah mammals. But there is not a *single* instance of this, not even among species like *baboons* and *vervets*, which are descended from forest-dwlling ancestors, but it leaves a lot of problems unanswered. On the question of why humans lost their *body hair*, for example, it has been argued at various times that no explanation is called for, or that we may never know the reason, or even that there may not be a reason. These attitudes seem to be not merely *defeatest*, but fundmentally *unscientific*.

The Aquatic Ape Theory (AAT) offers an *alternative* senerio. It suggests that when our ancestors moved onto the savannah they were already *different* from the apes; that n*akedness, bipedalism,* and other *modifications* had begun to evolve much *earlier*, when the *ape* and *human* first *diverged*.

AAT points out the most of the "enigmatic" *features* of human physiology, though r*are* or even *unique* among land mammals, are *common* in aquatic ones. If we postulate that our earliest ancestors had found themselves living for a prolonged period in a f*looded, semi-aquatic* habitat, most of the unsolved problems become much easier to unravel.

There is *powerful* geological evidence to *support* this hypothesis, and nothing in the fossil record is *inconsistent* with it. Some of the issues it *raises* are briefly outlined in the following pages.

The Naked Ape
Humans are classed *automatically* among the *primates*, the order of which includes apes, monkeys and lemurs. Among the hundreds of living primates species, only humans are n*aked*.

Two kinds of habitat are known to give rise to naked mammals – (1) a *subterraenian* one or (2) a *wet* one. There is a *naked* Somalian *mole-rat* which never ventures *above* ground. All other *non-human* mammals which have *lost* all or most of their fur are either s*wimmers* like *whales* and *dophins* and *walruses* and *manatees*, or wallowers like h*ippopotamuses* and *pigs* and *tapirs*. The *rhinoceros* and the *elephant*, though found on land since Africa became drier, bear traces of a more watery past and seize every opportunity of wallowing in *mud* or *water*.

It has been suggested that humans became *hairless* "to prevent overheating in the savannah". But no other mammal has ever resorted to this strategy. A covering of hair acts as a *defense* against the *heat* of the sun: that is why even the desert-dwelling camel r*etains* its fur. Another version is "to facilitate sweat-cooling". But again many species resort to *sweat-cooling* quite effectively withut *needing* to *lose* their hair.

There is no known reason why an ape should suffer *more* than from overheating than the savannah baboon. And, especially for a savannah primate, there would be a high price to pay for hairdressers. Primate infants are carried around *clinging* to their mothers' fur; the females would be severely hampered in their foraging when that no longer became possible.

One general conclusion seems *undeniable* from an *overall* survey of mammalian species: that while a coat of fur provides the *best* insulation for *land* mammals the *best* insulation in water is not *fur*, but a layer of *fat*.

Fat

Humans are by far the *fattest* primates; we have *ten times* as many fat cells in our bodies as would be *expected* in an animal of our size.

There are *two* kinds of animals which tend to acquire large deposits of fat – (1) *hibernating* ones and (2) *aquatic* ones. In hibernating mammals the fat is *seasonal*; in most aquatic ones, as in humans it is present all year round. In land mammals fur tends also to be stored *internally*, especially around the kidneys and intestines; in mammals and in humans a higher proportion is deposited *under* the skin.

It is unlikely that *early* man would have *evolved* this feature after *moving* to the plains and becoming a *hunter*, because it would have *slowed* him down. No land predator can a*fford* to get fat. Our *tendency* to put on fat is likelier to be *inheritance* from an *earlier* aquatic *phase* of our evolution. It is true that *some* apes, especially in *captivity*, may put on weight, but we still differ from them in *two* important ways. **One** is that they are never b*orn* fat. All infant primates except our own are *slender*; their lives may depend on their ability to *cling* to their mothers and support their whole weight with their *fingers*. Our own babies accumulate fat *before* birth and continue to grow fatter for several months a*fterwards*. Some of this fat is *white* fat, and that is certainly *rare* in new-born mammals. While fat is not much good for supplying *instant* heat and energy, it is good for *insulation* in *water*, and for giving *buoyancy*.

The **other** difference is that in our case the subterraenian fat is *bonded* in the skin. When an anatomist skins a cat or rabbit or chimpanzee, any superficial fat deposits remain *attached* to the *underlying* tissues. In the *case* of humans, the fat comes *away* with the *skin*, just as it does in *aquatic* species like *dolphins, seals, hippos* and *manatees*.

Walking on Two Legs

Human beings are the only mammals in the world that habitually walk on *two* legs – the only other creature with a perpendicular gait is the aquatic bird, the *penguin*.

It is not surprising bipedalism is so *rare*. Compared with running or walking on four legs it has many *disadvantages*. It is *slower*, it is relatively *unstable*; it is a skill that takes m*any* years to learn, and it *exposes* vulnerable organs to *attack*.

We have been doing it for **5 million years** and in that time our bodies have been drastically *remoulded* to make it *easier*, but it is still the *direct* cause cause of many discomforts and ailments such as *back pains, varicose veins, haemorrhoids, hernias* and problems in *childbirth*. It would have been far more difficult and laborious for our ape- like ancestors; only some powerful pressure could have induced them to adopt a way of walking for which they were *initially* so ill suited.

One hypothesis used to be that they *first* developed big brains and began to make t*ools*, and finally walked on their *hind* legs to *free* their hands for carrying weapons. But we now know that it was *bipedalism* that came *first*, **before** the **big brain** and **tool- making**.

If their *habitat* had become *flooded*, however, they would have been *forced* to walk on their *hind* legs whenever they came down to the ground in order to keep their heads a*bove* water. The *only* animal which has ever evolved a pelvis like ours, suitable for bipedalism, was the long-extinct *Oreopithecu*, known as the *swamp ape*.

Today, *two* primates when on the ground *stand* and *walk* erect somewhat more readily than most *other* species, include the *bonobo*, or *pigmy chimpanzee*, their habitat consisting of a tract of seasonally flooded forest, which would have covered an even more *extensive* area before the African climate became *drier*.

Both of these species *enjoy* the water. It is interestimg that the bonobos often *mate* f*ace-to-face* as humans do; in our case it is expected as a consequence of bipedalism. This mode of mating is another characteristic very *rare* among land animals, which we share with a *wide* range of aquatic mammals such as *dolphins, beavers*, and *sea otters*. What we have in *common* with them is a *mode* of locomotion in which the spine and the hind limbs are in a *straight* line, and that affects the *position* of the sex orgns.

Breathing

The human *respiratory* system is *unlike* any other land mammal's in two respects:

The **first** is that we have *conscious* control of our breathing. In most mammals these actions are *invountary*, like the *heart beat* or the processes of *digestion*.

Voluntary breath control appers to be an *aquatic* adaptation because, apart from ourselves, it is found *only* in *seals* and *dolphins*. When they decide how *deep* they are going to *dive*, they can *estimate* how much air they need to inhale. *Without* voluntary breath control it is very *unlikely* that we could have learned to **speak**.

The **other** human peculiarity is called "the descended larynx". A land mammal is normally obliged to breath through its nose *most* of the time, because its windpipe passes up thorugh the *back* of the throat and the **top** end of it (the larynx) is situated in the *back* of its nasal passages. A dog, for example, has to make a special *effort* to bring its larynx d*own* to the lungs in order to *bark* or to *pant*; when it *relaxes*, the larynx goes *back up* again. We are *born* with this.

A few months after birth the human larynx *descends* into the throat, right down below the back of the tongue. Darwin found that very *puzzling* because it means that the opening to the lungs lies *side-by-side* with the *opening* to the stomach. That is why in our species food and drink may sometimes go "down the wrong way". If we had *not* evolved an *elaborate* swallowing mechanism it would happen *every* time.

This arrangement means that we can breath through our mouths as easily as through our noses. It is probable that this is an *aquatic* adaptation, because a swimmer needing to gulp air quickly can inhale more of it through the *mouth* than through the *nostrils*. And we do know that *only* birds which are *obligatory* mouth breathers are diving birds like p*enguins, pelicans* and *gannets*. As for *mammals*, the only ones with a descended lartnx, apart from ourselves, are aquatic ones – the *sea lion* and the *dugong*.

Other Differences

It is impossible in a brief outline to discuss all of the physical features distinguishing **US** from the **apes**, but a *few* are worth mentioning:

We have, for example, a *different* way of *sweating* from other mammals, using different *skin glands*. It is very *wasteful* of the body's essential resources of *water* and *salt*. It is therefore *unlikely* that we acquired it on the *savannah*, where *water* and *salt* are both in *short* supply, or at times *not* available at all.

We *weep* tears of *emotion* by different nerves, from the ones that cause our eyes to *water* in response to smoke or dust. No other land animal does this. There are marine birds, marine reptiles and marine mammals which shed water through their eyes, or through special *nasal* glands, when they have *swallowed* too much seawater. This process may also be *triggered* in them by an *emotional* excitement *caused* by feeding or fighting or frustration. Weeping animals, apart from ourselves, include the *walrus*, the *seal* and the *sea otter*.

We have millions of *sebaceous glands* which exude oil over head, face and torso, and *in young* adults often causes *acne*. The chimpanzee's sebaceous glands are described as "vestigial" whereas ours are described as "enormous". Their purpose is obscure. In other animals the only function of sebum is that of waterproofing the skin or the fur.

The most widely discussed contrast between ourselves and the apes is that we have bigger brains. A bigger brain may well have been an advantage to a chimpanzee: the question is why one of them acquired it.

One factor may have been *nutritritional*. The building of brain tissue, unlike other body tissues, is dependent on an adequate supply of *Omeg-3* fatty acids, which are abundant in a marine food chain but relatively scarce in the land food chain.

AAT is the only theory which logically connects all these and other enigmatic features and relates to a single well established event.

The Time and the Place

It is now generally agreed that the man/ape split occurred in Africa between **7** and **5 million years ago**, during a period known as the *fossil gap*.

Before **it**, there was an *animal* which was the *common ancestor* of human and African apes. After **it**, there emerged a **creature** smaller than ourselves, but bearing the unmistakable hallmark of the first shift towards human status: it walked on *two legs*.

This poses two questions: (1) "Where were the earliest fossils found?" and (2) "do we know of anything happening in that place at that time that might have caused apes and humans to evolve along *separate* lines?"

The oldest pre-human fossils (including the best known one, "Lucy") are called *Australopithecus afarensis* because their bones were discovered in the *triangle*, and area of low lying land near the Red Sea, about **7 million years ago** that were flooded by the sea and became the *Sea of Afar*.

Part of ape population living there at the time would have found themselves living in a radically changed habitat. Some may have been marooned on off-shore islands – the present day Danakil Alps were *once* surrounded by water. Others may lived in *flooded* forests, salt marshes, mangrve swamps, lagoons or on the shores of the new sea, and they would all have has to adapt or die.

Chararacteristics	Humans	Apes	Savannah	Aquatics
Habitual Bipedalism	Yes	-	-	-
Loss of body hair	Yes	-	Yes	Yes
Skin-bonded fat deposits	Yes	-	-	Yes
Ventro-ventral copulation	Yes	Yes	-	Yes
Dimunition of apocrine glands	Yes	-	-	Yes
Hymen	Yes	-	-	Yes
Enlarged sebaceous glands	Yes	-	-	Yes
Psychic tears	Yes	-	-	Yes
Loss of vibrissae	Yes	-	-	Yes
Volitional breath control	Yes	-	-	Yes
Eccrine thermoregulation	Yes	-	-	Yes
Descended larynx	Yes	-	-	Yes

The "yes" in column 3 refers to the bonobo In column 4 the rhinoceros and the elephant

AAT suggests that some of them survived, and had begun to *adapt* to their watery environment. Much later, when the *Sea of Afar* became *landlocked* and finally *evaporated*, their descendents returned to the mainland of Africa and began to *migrate* southwards, following the waterways of the *Rift Valley* upstream. There is nothing in the fossil record to invalidate this scenerio, and much to sustain it. **Lucy's** bones were found at *Afar* lying among crocodile and turtle eggs and crab claws at the edge of a flood plain near what then have been lakes and rivers.

Other fossils of *Australopithecus*, dated *later*, were found further south, almost invariably in the immediate vicinity of ancient lakes and rivers.

We now know that the change from the *ape* into *Australopithecus* took place in a *short* space of time, by *evolutionary standards*. Such *rapid* speciation is almost invariably a sign that one population of a species has become *isolated* by a geographical barrier such as a stretch of **water**.

INTRODUCTION TO EVOLUTION UPDATES

Fear of Snakes Drove Pre-Human Evolution

An evolutionay *arms* race between *early* snakes and mammals triggered the development of *improved vision* and *large brains* in primates, a radical theory suggests.

This idea proposed by Lynn Isbell, an anthropologist at the *University of California, Davis* suggests that *snakes* and *primates* share a long and intimate history, one that forced both groups to *evolve* new strategies as each attempted to gain the *upper* hand.

To *avoid* becoming snake food, early mammals had to develop ways to detect and avoid the reptiles before they could strike. Some animals evolved better snake *sniffers*, while others developed *immunities* to serpent venum when it evolved. Early primates developed a better eye for *color, detail* and *movement* and the ability to see in *three- dimension* – traits that are important for detecting threats at *close* range.

Humans are Descended from these Primates

Scientists had previously thought that these *traits* evolved *together* as primates used their h*ands* and *eyes* to grab *insects*, or pick *fruit* or to *swing* through *trees*, but recent discoveries from *neuroscience* are casting *doubt* on these theories.

"Primates went a particular route," Isbell told *Live Science.* "They focused on i*mproving* their *vision* to keep away from [snakes]. Other mammals couldn't do that. Primates had the *pre-adaptiveness* to go that way."

Harry Greene, an evolutionary biologist and snake expert at *Cornell University* in New York, says Isbell's *new idea* is very exciting. "It strikes me as a very special piece of achievement and I think it's going to provoke a lot of *thought,*" Green said.

Isbell's work is detailed in the July issue of the *Journal of Human Evolution.*

A New Weapon

Fossil and DNA evidence suggests that the snakes were *already* around when the *first* mammals evolved some **100 million years ago.** The *reptiles* were thus among the *first serious* predators mammals faced. Today, the only *other* threats faced by primates are r*aptors*, such as *eagles* and *hawks*, and large *carnivores*, such as *bears*, large *cats* and w*olves*, but these animals evolved long *after* snakes.

Furthermore, these other predators can be safely detected from a *distance*. For snakes, the *opposite* is true.

"If you see them *close* to you, you still have *time* to *avoid* them," Isbell said. "Primate v*ision* is particularly *good* at *close* range."

Early snakes killed their prey using *surprise* attacks and by *suffocating* them to death, the method of boa constrictors. But the *improved* vision of primates, combined with other snake-coping strategies developed by other animals, *forced* snakes to evolve a new weapon: *venum*. This important milestone occurred about **60 million years ago**.

"The [snakes] had to do something to get *better* at finding their prey, so that's where v*enum* comes in," Isbell said. "The snakes *upped* the ante and then the primates had to respond by developing even *better* vision."

Since primates developed *vision* and *enlarged* brains, *three traits* became *useful* for other purposes, such as *social interaction* in *groups*.

Seeing in 3D

Isbell's *new* theory could explain how a number of primate-defining traits *evolved*.

Primates, for example, are among the *few* animals whose eyes face *forward* (most animals have eyes located on the *sides* of their heads). This so-called "orbital convergence" improved *depth* perception and allows monkeys and apes, including humans, to see in *three* dimensions. Primates also have better *color vision* than most animals and are also *unique* in relying *heavily* on *vision* when when *reaching* and g*rasping* for objects.

One of the most *popular* ideas for explaining how these traits *evolved* is called the "visual perception hypothesis." It purposes that our *early* ancestors were *small, insect mammals* and that the need *stalk* and *grab* insects at **close** range was the *driving* force behind the evolution of *improved* vision.

Another popular idea called the "leaping hypothesis," argues that orbital convergence is not only important for **3D** vision, but also for breaking through *camouflage*. Thus, it would have been *useful* not only for capturing insects and finding small fruit, but also for a*iming* at small, hard to see branches during mid-leaps through trees.

"But there are problems with *both* hypothesis," Isbell says.

First, there is no *solid* evidence that early primates were *committed* insectivores. It's possible that like many primates today, they were *generalists*, eating a *variety* of plant foods, such as *leaves, fruit* and *nectar*, as well as *insects*.

More *significantly*, recent studies do *not* support the idea that vision evolved *alongside* the ability to *reach* and *grasp*. The new *data*, rather, suggests that the *reaching* and *grasping* abilities of primates actually evolved *before* they learned how to *leap* and b*efore* they *developed* stereoscoptic, or **3D** vision.

Agents of Evolutionary Change

Isbell thinks proto-primates – the *early* mammals that eventually evolved into primates – were in a *better* position compared to *other* mammals to evolve *specialized* vision and e*nlarged* brains because of the *foods* they ate.

"They were eating foods *high* in *sugar*, and *glucose*, which is *required* for m*etabolizing* energy," Isbell said. "Vision is a part of the brain, and messing with brain takes a lot of *energy* so you're going to need a *diet* that allows you to do that."

Modern primates are among the most *frugvorous*, or "fruit-loving," of all mammals, and this trend might have started with the *proto-primates*. "Today there are *primates* that fo*cus* on *leaves* and *things* like that, but the *earliest* primates may have had a *generalized* diet that included f*ruits, nectar*, *flowers* and *insects*," she said.

Thus, *early* primates not *only* had a *good* incentive for developing *better* vision, they had already been eating *high-energy* foods needed to do so.

Testing the Theory

Isbell says her theory can be *tested.* Scientists, for example, could look at whether primates can visually detect snakes more *quickly* or more *reliably* than other mammals.

Scientists could also examine whether there are *differences* in the snake-detecting abilities of primates from around the world.

"You could see whether there is any *difference* between Malagasy lemurs. South American primates, and the African and Asian primates," she said. "Anthropologists have tended to *stress* thngs like *hunting* to explain the special a*daptations* of primates, and particularly *humans*," said Greene, the *Cornell* snake expert, but scientists are starting to *warm* to the idea that *predators* likely played a *large role* in h*uman evolution* as well.

"Getting *away* from *things* is a big deal, too," Greene added. "Snakes and people have had a *long* history: it goes back to *long before* we were people in fact," he said. "That might sort of explain why we have such *extreme* attitudes towards snakes, varying from d*eification* to "ophdipiphobia," or *fear of snakes*.

Were Bacteria the First Forms of Life on Earth?

For as long as we have known about them, scientists have thought that simple *bacteria* were the *link* to the *earliest* forms on earth. Growing *evidence*, however, suggests we've got it all *backwards* – could the secrets of the origins of life be *lurking* **inside** us?

When *most* of us think about evolution, we tend to think in terms of *simple* organisms evolving into *more* complex ones. Simple chemical reactions *evolved* into *simple* cells, which later *evolved* into more *complex* organisms, and so on all the way *up* to humans.

It's no longer believed that humans are at the *top* of the evolutionary ladder, but evolution does tend to drive organisms towards *greater* complexity.

This is not, however, *always* so. Rather, these organisms that leave the *most* offspring behind, simple or complex, do *best*. Greater complexity is sometimes a cansequence of evolution, but simplification can also be a *winning* strategy – It all dpends on the e*nvironment*. Most scientists nevertheless hold that the first organisms on earth were much *like* bacteria of *today*. But *several* features of the biochemistry of life suggest that bacteria aren't so *ancient* after all. In some respects, actually, the *cells* of our *own* bodies tell us more about the *evolution* of life than bacteria do. The key is in the discovery that won *Sidney Altman* and *Tom Cech* the **Nobel Prize** in *Chemistry* in 1989.

The Chicken and the Egg

1. In modern organisms, genetic information is *stored* in DNA (deoxyribose nucleic acid) in units called *genes*.

2. Genes code for *proteins*, which are responsible for the various *activities* that make a cell function.

3. Some proteins, called *enzymes*, perform the chemical reactions that *run* the cell.

   One *group* of enzymes, the DNA polymerases, make DNA.

4. The *information* for making those enzymes is *stored* in the DNA as a **gene**.

Therefore, *finding* the evolutionary origin of proteins and DNA is *tricky* as each r*equires* the *other* for its own *synthesis* – which came first? That's where Cech and Altman come in. They studied RNA (ribose nucleic acid), a *close* chemical relative of DNA.

1. When proteins are made from the information in DNA, a *working* RNA *copy* of the gene is made for use by *ribosomes* – the *protein factories* of the cell. RNA therefore, like DNA, *stores* genetic information, and, like proteins, it *also* performs chemical reactions.

2. Bringing RNA into the picture *solves* the *chicken* and *egg* problem, RNA can be *both chicken* and *egg*.

3. What this *means* to evolutionary biologists is that LIFE could have BEGUN with o*rganisms* made largely of RNA.

4. This idea of an 'RNA world' has been debated since the 1900s, but Chec and Altman's discovery has convinced most scientists that it is at least possible.

5. It is now *known* that RNA is at the *heart* of many of the *basic* functions in the cell, and probably evolved in the RNA world.

Jack of All Trades But if RNA is so *versatile*, what *happened* to the 'RNA world?' As the saying goes, "Jack of all-trades, but master of none," RNA is *not* as good at *performing* chemical reactions as *proteins*, nor is it as good at *storing* genetic information as DNA. It's not surprising that since the **RNA** world, **proteins** have gradually *replaced* most RNA e*nzymes*, and ***DNA*** *now* stores the *genetic* information.

Most *researchers* agree that RNA *used* to have *more* of a *central* role because it s*olved* the *chicken-and-egg* problem of which came *first* in evolution – proteins or DNA.

But other than being a tidy piece of logic, is there any *substance* to the hypothesis? Since we *can't* travel back in time, it is *impossible* to *prove* the existence of the RNA world outright, but we can do the next best thing – *rebuild* it from the '**molecular fossils**' that have *remained* in modern cells.

Looking For 'Molecular Fossils'

By deleving into our RNA-rich past, we can get an idea of what *early* life looked like, and this may in turn help us understand *how* life *evolved* into the *many* forms we see today. 'Digging' for molecular *fossils* is *not* a *trivial* exercise. Not all RNAs are going to be *genuine* RNA world fossils, but it is possible to *establish* which are likely to be a*ncient*, and which are more *recent* additions.

1. One of the most *central* machines in the cell is the *ribosome,* which translates the genetic code stored in DNA into the *language* of proteins.
2. The *core* of the *ribosome* is made from RNA, with *proteins* providing a *scaffold* to h*old* the RNA in place. Even with *most* of the RNA *stripped* away, it can still *make* proteins.

3. The *finding* that the RNA *core* is the *engine* room of the protein synthesis factory is a s*trong* argument that *protein synthesis* was *invented* in the RNA world.

Indeed, the *ribosome* is just *one* of *many* RNA machines, giving scientists a surprisingly *clear* picture of *early* life. The picture appears *most* complete in the e*ukaryotes* (plants, animals, fungi, amoebae), which are more *reliant* on RNA than p*rokaryotes* (bacteria and related cells), the *former* retainng more *clues* to *our* RNA-rich past. Since RNA is "a jack-of-all-trades" we would expect it to be gradually *replaced* during evolution.

The *common* view of the *evolution* of *eukaryotes* and *prokaryotes* from the ancient 'last *universal common ancestor*' (the LUCY), is that a prokaryote-like *creature* was the f*irst* to arise, and that an *ancient* eukaryotes arose from some prokaryote:

RNA world to LUCY to prokaryote-like organism to eukaryote-like organism Since many *more* ancient RNA world fossils are *found* in *eukaryotes*, the graph makes more sense this way, with a gradual *loss* of RNA.

RNA world to LUCY to eukaryote-like organism to prokaryote-like organism.

Backing Up the Hard Drive

In *addition* to the *information* we can glean from the study of RNA, we can also learn about our *evolutionary past* by examining *how* the *storage* of *genetic* information has ev*olved*. Genetic information is like any other information; it is *stored* in a particular medium, it gets *copied, read* and *transmitted*, and over time, small *errors* can turn up.
Prokaryotes and *eukaryotes* employ quite *different* mechanisms for *ensuring* that genetic information is *not corrupted*. The differences as follows:

1. Prokaryotes maintain their genome in a *single* cell, usually a *single* DNA chromosome, with as *much* information *packed* onto it as possible.

2. Many *eukaryotes* have *two* copies, *divide* the information up into *several* DNA chromosomes, with the genes packed relatively *sparsely*. Unlike prokaryotes, eukaryotes keep *backup* copy for its rescue.

3. Having *only* one copy of a gene means that if an *error* is made, that *damage* is p*ermanent* – there is no *backup* copy for its rescue.

   But what *difference* does it make if genes are split up or kept on a *single* chromosome?

Consider this example:

1. *Genome 1* also has *two* genes, *A* and *B*. These are housed on a *single* chromosome, and there are *two* copies of that chromosome. In one copy, gene *A* gets damaged and in the other, gene *B* gets damaged. The organism can *still* survive because it still has o*ne* working copy of gene *A* and one working copy of gene *B*. It, however, has to c*arry* the damaged copies also, because these are *physically* linked (on the same chromosome) to the working copies.

2. *Genome 2* also has *two* genes, *A* and *B*, each gene is housed on its *own* chromosome, and again there are *two* copies of each chromosome. Again, one copy of gene *A* gets damaged, and *one* copy of gene *B* gets damaged. *Genome 2*, however, can *discard* the damaged copies of genes *A* and *B* without *losing* the *working* copies, as *each* gene has its own chromosome.

Genome 'architecture' gives us a *clue* as to the informational stresses on a genome: the m*ore* genes that are *kept* together on a *single* chromosome, the *greater* the accuracy required to maintain them.

1. In *early* genetic systems, the strategy described for *genome 2* was probably used because damage was *frequent*, and thus provided a good way of *discarding* damaged genes while *keeping* the undamaged copies.

2. The *more* copies of a genome and the *fewer* genes per chromosome, the *better* this would work

3. *Prokaryotes* appear to use *none* of these safety nets, so can be considered *risk takers.*

4. *Prokaryotic* genomes thus look like a *recent* invention-organisms and could afford to keep *only one* copy of their infomation on a *single* chromosome, if they were pretty *sure* they would *not* lose it.

5. *Early* life forms were *not* good at *storing* information, especially since they had RNA as a *genetic* material. They needed to *develop* as many little *tricks* as possible to m*inimize* mutation; genomes like those in *prokaryotes* would have been disastrous.

6. Perhaps *eukaryotes* (plants and animals), for example, never *lost* many of those ancient *traits*, left over from a period where *copying* was *poorer*, so their genomes can be considered *fossils* of this *earlier* period in the evolution of life.

7. Interestingly, many single-celled *eukaryotes maintain* their DNA as a *single* copy. It is therefore *hard* to know if ancestral *eukaryotes* maintained *one* or *more* copies of their genome.

8. Keeping *back-up* copies was nevertheless an essential *trait* in very *early* genetic systems.

The Tortoise & the Hare

Looking at *both* the RNA *relics* in *eukaryotes* and their genome *design*, it appears e*ukaryotes* have *maintained* the status quo, *retaining* many molecular fossils, while p*rokaryotes* have *lost* many of these. How might this have happened? It could well be due to 'lifestyle' – the ways in which organisms go about their *daily* business.

1. Some organisms (e.g., oak trees) are *slow* growing and *rely* on a stable supply of n*utrients*, making their populations fairly *stable* also.

2. Other organisms (e.g., locusts) *grow* very *fast* and compete for nutrients that are extremely variable in supply. When a nutrient is *available*, it is *crucial* to grow and reproduce as *fast* as possible. When the food supply runs *low*, the populations suffer large crashes and only a *few* make it to the next source.

For some organisms, *speed* is everything – the ability to *react* quickly to the presence of a *new* food source is all-important. This means that if one organism is *faster* to r*espond* than the rest, it will *profit* at the *expense* of its competitors. Thinking back to the a*ncient* RNA machinery, there would have been strong *selection* in organisms to *replace* in*efficient* RNA machinery with *faster* protein machinery.

In *modern* organisms, *eukaryotes as* a group *fit* into the 'Tortoise' group, while the p*rokaryotes* take the *fast* 'hare' track, but within these, there is *also* a spectrum-brewer's yeast that has a 'bacterial lifestyle' when compared to oak trees for example. Even *fast* growing *eukaryotes*, however, have a lot of *slow* RNA machinery, so this argument alone may *not* explain how *prokaryotes* arose. If a fast lifestyle alone doesn't *shed* RNA, then how do we explain the *lack* of RNA in *prokaryotes*?

Shedding the Excess in Life's Sauna

The evolutionary 'push' that gave *rise* to the *prokaryotes* may well have been an adaptation to living at *high* temperatures. Patrick Forterre, at the *University of Paris* has put forth a hypothesis which he calls the 'thermoreduction hypothesis.' He maintains that p*rokaryotes* arose from a *eukaryote-like* ancestor by *adaptation* to life at *high temperatures*, and in the process *shed* many of their *ancient* features. Porterre's work suggests that even the bacteria now living at *moderate* temperatures retain traces of their h*ot history* (I find this **hot history** hypothesis *very* logical & convincing– Author's note).

Forterre's argument rests on the obervation that RNA is very unstable at high temperatures:

1. Organisms living at *high temperatures* should make *limited* use of RNA.

2. In *prokaryotes*, many RNA *fossils* appear to have been *replaced.*

3. Numerous *prokaryotes* live in the *scalding* hot temperatures typical of hot springs and deep sea thermal vents, which often *exceed* **100 degree centigrade**.

4. In *eukaryotes*, many RNAs are still in use.

5. Currently there are no *known* examples of *eukaryotes* living at extreme temperatures.

If any are found, we would expect them to *have* a lot of their RNA.

Another piece of evidence for *thermoreduction* comes from the genomes of p*rokaryotes*:

1. In *eukaryotes*, chromosomes are made of *linear* DNA.

2. In *prokaryotes* the genome is made of *circular* DNA.

3. *Circular* DNA is much *less* vulnerable to heat *damage* than *linear* DNA, which starts to get 'split ends' at *high temperatures.*

4. *Circular* chromosomes are *conspiculously absent* from *eukaryotes* and their widespread incidence in *prokaryotes* alone is *best* explained by the *thermoreduction* hypothesis (The fact that the earth was still relatively *young* and *hot* **3.5 billion ago**, when *life* first started, *lends* additional *support* for this *hypothesis* – Author's note).

For *eukaryotes* to maintain *linear* DNA genomes, they *require* a *special* system for maintaining their ends:

1. An enzyme called *telomerase*, which has *both* a *protein* and an RNA *component*, does this job.

2. *Telomrese* is *common* to all *eukaryotes*, suggesting it is *very ancient.*

3. It seems *unlikely* that *eukaryotes* with *linear* genome, and *many* RNAs, including t*elomerase*, could have emerged from the 'sauna' of life.

4. It is more likely that the organisms which first *braved* high temperatures *shed* much of the *evidence* of their RNA world *ancestry* along the way, as well as *linear* DNA g*enomes* and *telomerase.*

5. *Modern prokaryotes* appear to have had a 'hot history,' even though many now live at m*oderate* temperatures.

Clues to the Origin of Life in Your Own Body

Evolutionary biologists have traditionally *studied* the *simpliest* organisms they can find in order to learn *more* about the **origins of life**. But *simple* doesn't necessarily mean a*ncient*, so we should not *restrict* our search *purely* to *simple organisms*. All organisms have been *evolving* for **3.5 billion years or so**, and the idea that there is some *obscure* bug that time *forgot* which *resembles* ancient life is *outdated*.

As Forterre's work shows, *simplification* has its *merits*, and it seems that bacteria have l*ost* a lot of molecular *fossil* of our *ancient* past. We know an *enormous* amount about the biochemistry of our *cells*, and although there's layer upon layer of completely, *hidden* underneath it all are *clues* to the *origins* of the *earliest* cells. How *ironic* it is that human cells *harbor* as many if not *more* secrets on the *origins* of life than the *simple* bacteria! It's *no* wonder that evolutionary biologists are as *excited* about the **Human Genome**

Project as anyone else!

It is important to keep in mind that *eukaryotic* cells have continued to *evolve* over time. While it is possible to uncover much about the RNA world, and how *prokaryotes* and *eukaryotes* evolved by *looking* at 'molecular fossils,' most features of the *eukaryotic* cell are 'recent' *innovations*. *Mitchondria* (the *power plant* of the *eukaryotic* cell), and c*hloroplasts* (the *organelle* which sunlight turns to *sugar* in plants), for instance, are the r*emnants* of ancient *prokaryotes* that were *engulfed* by ancient *eukaryotes*. Another major innovation is the *evolution* of *multicellular* organisms. This *brought* with it the fruits of d*ivision* of labor, allowing the evolution of *complex* organs and tissues.

Most researchers would *add* the *nucleus* to the list of 'new' *eukaryotic* features, but it's interesting to note that *all* RNA world *fossils* are *found* in the *nucleus*. The asuumtion that the *nucleus* is *recent* is *based* on the *argument* that evolution *drives* towards c*omplexity*, but we know this *isn't* always so. It is *exciting, though,* to consider the possibility that the *nucleus* is *old*, and *prokaryotes* have *lost* it.

Once upon a time, we'd probably have been in *danger* of being *burnt* at the *stake* for such *heretical* stuff, but nowadays, however, biologists no longer *view* the evolution of life as a *progression* from *simple* to *complex* with *humans* as the *pinnacle* of evolutionary achievement. Here's to your **molecular fossils!!!**

New Fossil: Link Between Fish and Land Animal

What creature *first* crawled out of the prehistoric swamps to *conquer* the land? The question has long *puzzled* paleoanthropologists because the *transitional* species seems to have lived during the *mysterious* **30-million-year gap** in the fossil record called **Romer's Gap**.

Now a rsearcher in Britain has *found* a very *rare* fossil of a short, squat crocodile-like c*reature* that she believes provides a *stepping stone* between our *aquatic* ancestors and the *first* four-legged land dwellers.

"This really is one of a find," said Jennifer Clark, a paleoanthropologist at the *University Museum of Zoology* ***Cambridge, United Kingdom.*** "It may even be the ***first*** five-toed foot," she added, "but I can't swear to that ***yet***."

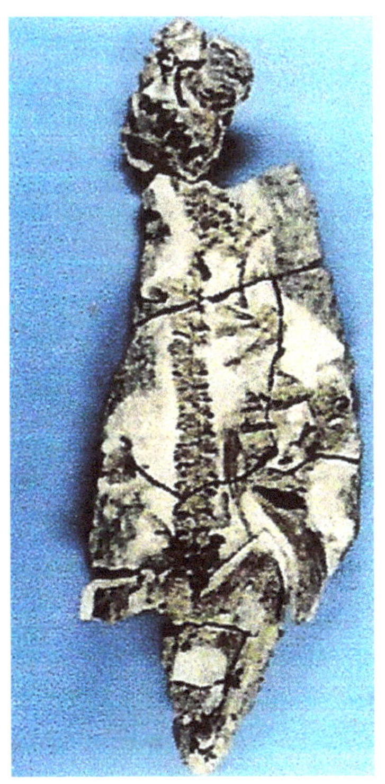

Pederpes

Shallow Lifestyle

Pederpes probably came from a *shallow-water* environment – a lagoon or coastal flat – that was vulnerable to *sudden* increases in salinity as water levels rose and fell. *Romer* believed such an environment would have favored the evolution of *limbs* enabling land travel, to allow the animal a *broader* area to feed. The discovery is published in the July 4 issue of the Journal *Nature*.

Clark feels that *early* evolution of hands and feet first occurred in relationship to strictly *aquatic* locomotion, somewhat in the manner of seals.

The fossil was *originally* discovered in 1971 in a stream that cuts through moorland about three kilometers from Dumbarton, New Scotland, and lay incorrectly classified in the *Hunterian Museum,* in Glassgow. The skeleton is almost *complete*, just missing a *tail*, and is thought to date back to **348 million years ago** – the *heart* of *Romer's Gap*. The *Gap* is named for *Alfred Sherwood Romer*, an American paleoanthropologist and a *prolific* writer of *textbooks* in the **1955s,** who *first* recognized the *lack* of fossils from this **30-million-year period**.

New Feet

The hint that Clark's *new* species, called **Pederpes finneyae**, might provide a *missing link* between the *swimmers* and the *landlubbers* lies in the *bone structure* of the *hind* leg.

During the *Late Devonian Period,* about **365 million years ago**, *tetrapods* had paddle- like feet for swimming that pointed *back* to the *side*, said Clark.

But *Pederpes* Feet are Different

"*Pederpes* looks as if its *feet* have been *reoriented* to *point* forward – perfect for *locomotion* on land," said Clark. That is, the *middle* toe on each foot points *straight ahead* just as it does in *modern* tetrapods like *dogs, mice and humans*.

The *Late Devonian Period* has a *rich* fossil history of *fin-lobed* fishes. The *earliest* four-legged specimens like *Acanthostega* and *Ichthyostega* lived about **363 million years ago**. *Acanthostega* had *limbs* and *eight digits* on each hand and foot, and also had fish characteristics like *gills, fins*, and *sensory organs* that worked *only* underwater. But these *fishy* animals probably *rarely* left the water, said Per Ahlberg, a paleoanthropologist of *fossil fish* and *amphibians* at the *National History Museum* in London.

After the *Devonian* the fossil record *diappears*, at least for a while – **20-30 million years**. Only three *informative* fossils *dating back* to this *time* have been *found.*

After Romer's Gap

When the fossil record *resumes* roughly **25 million years** later, there was already a tremendous *variety* of tetrapod landforms. *Ancestors* of modern mammals, amphibians, reptiles, and birds had *akready* evolved and were diverging along *distinct* branches.

And that left questions:

"We *lack* a *focus* from which *all* modern tetrapods *evolved*," said Robert Carrol, professot of zoology, curator of vertebrate paleontology at Montreal's *McGill University*.

"Romer's Gap is a **30-million year** black box that, frankly, keeps me *up* at night."

Clark's Latest Find May Help Scientists Better

Pederpes will be particularly *useful* for the purpose of "recirproal illumination," said John Bolt, curator of fossil amphibians and reptiles at *The Field Museum* in Chicago. "Seeing a *new* complete skeleton *adjusts* our mindsets to see *new* features in *fossils* that have *already* been examined. Otherwise we often see *only* what we *expect* to see."

The **Pederpes** fossil was originally *misclassified* as a *rhizodont* – an extinct type of *lobed* fish with the equivalent of upper arm and big bones. In the **mid-1990s**, one of Clark's graduate students, *persuing* the collection at the *Hunterian Museum* noticed the fossil – basically a *lump* of rock with a few teeth protruding – and brought it back to *Cambridge* for *further* study.

"When I saw the ***rock*** I became excited. I could see some ***scales*** covered the ***belly*** which were quite ***unlike*** fish scales," Clark said. "I also noticed ***another*** protruding bone called the ***ischium*** and I became very excited because I knew that this was ***not*** a fish but a tetrapod!"

Lessons From The Orangutans: Upright Walking May Have Begun In The Trees By observing *wild* **orangutans**, a research team has found that walking on *two* legs may have arisen in relatively *ancient*, tree-dwelling apes, rather than in more *recent* human ancestors that had already *descended* to the savannah, as current theory suggests.

Upright walking, or *bipedalism* has long been considered a *defining* feature of humans and our *closest* ancestors. One of the most *popular* explanations, known as the **savannah hypothesis**, suggests that the ancestors to chimps, gorillas and humans descended from the *trees* and began walking on the *ground* on all fours.

From All-Fours to Knuckle to Upright Walking

Over time, this four- legged *gait* would have e*volved* into the "knuckle-walking" that c*himps* and *gorillas* still u*se* today and then into u*pright*, two-legged walking in humans.

Paleotologists have conventually used *signs* of bipedaslism as *key* criteria for distinguishing *early* human, or "hominin," fossils from those *other* apes. But, this distinction is *complicated* by *recent* fossil evidence that some *early* hominins, including **Lucy** (Australopithecus afarensis), lived in *woodland* environments, while even *earlier* forms such as *Millennium Man* (**Orrorin**) appear to have *lived* in the forest canopy and moved on *two* legs.

"Our findings *blur* the picture even further," said Robin Crompton of the *University of Liverpool* in Liverpool, Great Britain, who is now *one* of the study's authors. "If we're right, it means you can't *rely* on bipedalism to tell you whether you're looking at a h*uman* or other *ape* ancestor. It's been getting more and more *difficult* for us to say w*hat's* a human and *what's* an ape, and our work makes that much more the case.

Crompton and his colleagues, Susannah Thrope and Roger Holder of the *University of Birmingham,* Great Britain, came to their conclusions by *observing* wild orangutans in Sumatra, Indonesia. Orangutans *spend* almost their whole lives in *trees*, making them useful *models* for how *our* ancestors *moved* around **several million years ago**.

To *collect* the data, Thorpe spent a year *living* in the Sumatrian rainforest and r*ecording* virtually every *move* the orangutans made. Then, she and her colleagues used these observations to *test* the hypothesis that bipedalism would have *benefited* our tree- dwelling ape ancestors.

Because these ancestors were probably *fruit-eaters*, as orangutans are, they would have needed a *way* to navigate the thin, flexible branches in the tree's periphery, where the fruit typically is. Moving on *two* legs and using their *arms* primarily for balance, or "hand-assisted bipedalism," may have helped them *travel* on these branches. The researchers *analyzed* nearly **3,000** examples of observed orangutan movement, and found that the orangutans were more *likely* to use hand-assisted bipedalism when they were on the *thinnest* branches. When *bipedal*, the animals also tended to grip *multiple* branches with their long toes.

On *medium-sized branches*, the orangutans used their arms *more* to support their w*eight*, changing their moving style to incorporate hanging. They only tended to walk on a*ll fours* with their *toes* and thus *distributed* their *center of gravity* more effectively while keeping one or both of their long arms *free* to reach for fruits and other supports.

Orangutans also keep their legs *straight* while *standing* on *bending* branches, the authors report. The exact benefit of the straight legs is still *unclear*, but when humans run on springy surfaces, they also keep their weight-bearing relatively straight, so this may have an *energy-related* advantage.

"Our results suggest that bipedapism is used to navigate the *smallest* branches where the *tastiest* fruits are, and also reach further to help to *cross* gaps between trees," said Thorpe.

Sumatrian Orangutan

The authors *propose* an evolutionary scenerio that begins as other researchers have envisioned. Somewhere toward the end of the Miocene Epoch (**24 to 5 million years ago**), climate in East and Central Africa became alternately *wetter* and *drier*, and the rainforest grew increasingly *patchy*. Apes living in the forest canopy would have begun to encounter *gaps* between trees that they could not c*ross* at the canopy level. The science authors suggest that early human ancestors r*esponded* to this by *abandoning* the high canopy for the forest floor, where they r*emained* bipedal and began eating food from the *ground* or *smaller* trees. The ancestors of chimps and gorillas, on the other hand, became more *specialized* for *vertical* climbing between the high canopy and the ground and this developed *knuckle-walking* for *crossing* from one tree to another on the ground.

"Our *conclusion* is that *arboreal* bipedalism had *very* strong *adaptive* benefits. So, we d*on't* need to explain how our ancestors could have gone from being *quadrupedal* to being *bipedal*," Thorpe said.

Observations of orangutan movement should be *useful* for conservation efforts, according to Thorpe. These animals are seriously *endangered*, primarily due to *habitat* destruction.

"If you can understand how they *cross* gaps in the forest, you can learn about *effects* that living in *logged* or *degraded* habitat would have on their locomotion. *These* could affect *energy* levels, for example, if they have to go to the *ground*, which *is* incredibly r*isky* because the Sumatran *tiger* is down there *licking* its lips. The Sumatran orangutan population is predicted to be *extinct* in the next *decade* if habitat degedation continues. Our research further highlights the *need* for protecting these animals," she said.

Walk the Walk

A *shape* comparison of the most *complete* fossil *femur* (thigh bone) of one of the *earliest* known *pre-humans*, or *hominins* (if you will), with the *femora* of living *apes*, modern h*umans* and other *fossils*, indicates the *earliest* form of *bipedalism* occurred at least **six million years ago** and *persisted* for at least **four million years**. Wilson Jurgers, Ph.D,. of *Stony Brook University*, and Brian Richmond, Ph.D., of *George Washington University*, say their finding indicates that the *fossil* belongs to very *early* human ancestors, and that u*pright* walking is one of the *first* human characteristics to appear in our lineage, right after the *split* between *humans* and *chimpanzees*. Their findings are published in the **March 21** issue of the *Journal Science*.

The research is the first *thorough* quantitive analysis of the *Orrorin* tugernensis fossil, a fragmentary piece of *femur* – was *discovered* in Kenya in **2000** by a French research team. Dr.Jurgers, Chair of Anatomical Sciences at SBU *School of Medicine*, and Dr. Richmond, Associate Professoor of Anthropology at GWU's *Center for the Advanced Study of Hominid Paleobiology* completed a *multivariate* analysis of the proximal femora s*hape* of a *young adult* Orrorin. *fugernensis* that enabled them to pinpoint the *pattern* of bipedal *gait* for this *controversial* hominin. Their analysis included *Astralopithecus*, and modern *humans* of *all* body sizes.

"This research *solidifies* the evidence that the human lineage *split off* as far back as **six million years ago**, that we *share ancestry* with **Orrorin**, and that our ancestors were walking *upright* at the time," says Dr. Richmond. "These answers were not *clear* before this analysis.

"Our *study* confirms that as early as **six million years ago**, *basal* hominins in Africa were already *similar* to later *australopthecus* in their *anatomy* and *inferred* locomotor biomechanics," adds Dr. Jurgers. At the same time, by way of the analysis, we see *no* special phylogenetic connection between *Orrorin* and *our* own genus, *Homo*.

In *Orrorin fugenensis Famous Morphology and the Evolution of Human Bipdalism*, the author's *articulate* that the analysis and morphological *comparisons* among *femora* from the *fossils* showed that *Orrorin fugenensis* is *distinct* from those of modern h*umans* and the *great apes* in having a long, anteroposteriorly narrow *neck* and wide proimal shaft. *Early* Homo *femora* have larger *heads* and broader *necks* compared to early *hominins*. In addition to these features, modern human *femora* have short *necks* and mediolaterally narrow *shafts*.

The *challenge* ahead, stipulates Dr. Jurgers, is "to identify what precipitated the c*hange* from the *ancient* and successful *adaptation* of *upright walking*, and *climbing*, to our *own* oblique form of *bipedalism*."

Origin of Bipedalism Seems Most Closely Tied To Environment During the past **100 years**, scientists have *tossed* around a great many hypothesis about the evolutionary *route* to *bipedalism*, and what *inspired* our *prehuman* ancestors to stand up straight and *amble* off on two feet.

Now after an *extensive study* of evolutionary, anatomical and fossil evidence, a team of paleoanthropologists has *narrowed* down the number of *tenable* hyothesis to explan the *origin* of bipedalism and our *prehuman* ancestors' method of navigating their world before they began walking upright.

The hypothesis they found the *most* support for regarding the *origin* of bipedalism is the one that argues our ancestor began walking upright largely in *response* to e*nvironmental* changes – *in particular*, to the growing *incidence* of *open* spaces and the way that *chaned* the distribution of *food*. In response to periods of *cooling* and *drying*, which *thinned* out dense forests and produced "mosaics" of *forests, woodlands* and g*rasslands*, it seems *likely* that "some apes *maintained* a forest-oriented *adaptation*, while o*thers* may have begun to *exploit* forest margins and grassy woodlands," said paleoanthropogist Brian Richwood, *lead* author in the *new* study. The process of increasing commitment to bipedality probably "an extended and complex o*pening* of habitats, rather than a *single*, abrupt *transition* from *dense* forest to *open* savannah, " he said.

Richmond, from the *University of Illinois at Urnana-Champaign*, with paleoanthrpologist David Strait from the *New York College of Osteopathic Medicine*, describe their findings, which involved a comprehensive review and analysis of the *five* l*eading* hypothesis on the *origin* of bipedalism, in a recent issue of the *Yearbook of Physical Anthropology*. Other hypothesis that remain *viable*, according to the team," are *freeing* the hands for carrying or for some kind of tool use, and an increased emphesis on *foraging* from the branches of small fruit trees, remains in the context in which modern chimpanzees *spend* the most time on *two legs*.

For their *study*, the researchers *combined* data from *biomechanics – movement* and s*tresses* in *bones* and *joints* and from bone *growth* and *development*. They *found* that our prehuman ancestors had *terrestrial* features in the *hands* and *feet*, climbing *features* and throughout the skeleton, and **knuckle-walking** *features* in the **wrist and hand**; that finger bone *curvature* is responsive to *changes* in arboreal activity during growth, lending support to the hypothesis that many *early* hominin species, although bipedal, still *climbed* trees. Evidence from the wrist joint "suggests that the earliest humans evolved bipedalism from an ancestor *adapted* for knuckle-walking on the *ground* and *climbing* in trees." said the researchers.

The YPA article, according to Richmond, is the *first* attempt in decades to bring t*ogether* all of the available evidence for the argument that the *earliest* human bipedalism e*volved* from *ancestors* that both *knuckle-walked* and *climbed trees*, rather *than* from ancestors *living exclusively* in trees and coming *down* from their *trees*.

Infant Carrying Ruled Out As Reason Why Early Humans

Walked Upright, According To New Research

Scientists investigating the reasons why *early* humans – the so-called *hominins* – began walking *upright* say it's *unlikely* that the need to *carry* children was a *factor* as has p*reviously* been suggested.

Carrying babies that could no longer *use* their feet to *cling* on to their parents in the way that young apes can has long been thought to be at least *one* explanation as to why humans became bipedal.

But *University of Manchester* researchers investigating the *energy* involved in c*arrying* a child say the physical *expense* to the mother does not *support* the idea that walking upright was an evolutionary *response* to child transportation.

"Walking upright is one of the **major** characteristics that **separates** humans from their primate relatives," said Dr. Jo Watson, who carried out the research in the University's *Faculty of Life Science*.

"Scientists have long hypothesized as to the reasons why hominins became bipedal in such a relatively *short* space of time but the *truth is* we still *don't* know for sure."

"One of the most *popular* explanations is that walking upright *freed* our forelimbs allowing us to *carry* objects, including children; modern apes, however, have *no* need to carry their young as they are able to *grip* their young using *both* hands in *each* instance.

The team *monotered* the *oxygen* consumption of seven women, all *healthy* individuals u*nder* the age of 30, carrying *either* a *symmetric* load, in the team of a weighted vest or a **5kg** dumbbell in *each* hand, or an *asymmetric* load, which was a single **10kg** weight carried in *one* hand or mannequin infant on *one* hip.

"Carrying an awkward *asymmetric* load, such as an *infant* on one side of the body, is the *most energetically* expensive way of *transporting* the weight," said Dr. Watson, whose research is published in the *Journal of Human Evolution*.

"Unless infant carrying resulted in significant benefits elsewhere, the *high* cost of carrying an *asymmetrical* weight suggests that infant carrying was *unlikely* to have been the evolutionary *driving force* behind *bipedalism,*" said Dr. Watson.

The study, carried out with *colleagues* at the *Universitys of Sheffield* and *Salford* and funded by the *Natural Environment Research Council* (NRIC), is *part* of a larger project, run by Dr. Bill Sellers at the *University of Manchester*, which uses *computer simulations* to understand evolutionary processes, *particularly* the way in which we and other animals m*ove*.

Future plans are to *extend* this work to *assess* the *energy cost* of carrying in great apes. Computer *models* of *early* hominins carrying loads will will also be built to try and e*valuate* whether their body shape and posture – long arms and short legs – would have made them *noticeably* better or worse at carrying *than* present-day humans. The research team hopes this will help build up a *picture* of how humans *evolved* to *walk* on *two legs*.

Human Ancestors Went *Out Of Africa* And Then *Came Back*

SUNY-Albany biologist Caro-Beth Stewert and NYU anthropologist Todd H. Disotell have *proposed* a controversial *new model* for the evolution of *humans* and *apes*, which t*ogether* are called the *hominoids*. Stewert and Disotell argue that the *ancestor* of humans and the *living* African apes *evolved* in *Eurasia*, not *Africa*.

This controversial *new* model for the evolution of humans and the apes is the cover story of the **July 30th** issue of *Current Biology*. Stewert and Disotell describe their theory in the article entitled "*Primate evolution – in and out of Africa*."

Today, the *lesser apes* (gibbons and siamanga) and some *great apes* (orangutans) *live* in Southwestern Asia, while other *great apes* (gorillas and chimpanzees) live in Equatorial Africa. The *fossil record* indicates that apes were present in Europe and Western Asia during the *Miocene Era*, from about **8 to 17 million years ago**. Ancestors of these ape species must have *moved* between the *African* and *European* land masses during their evolutionary *history*. According to the theory traditionally *held* by most paleoanthropologists, the *hominoids* evolved in *Africa*. The *lesser apes* and *orangutans* subsequently *dispersed* out of *Africa* to *Eurasia* at different times, leaving behind representatives of the lineage leading to the gorillas, chimpanzees and humans.

Based on a *synthetic* analysis of molecular, fossil, and biogeographical data for the primates, Stewert and Disotell propose *instead* that the lineage leading to the *common* ancestor of all *living* apes *dispersed* out of Africa about **20 million years ago** (during the early *Miocene*) and then *speciated* into the *greater* and *lesser ape lineage* in Eurasia. Within the past **10 million years**, one of the great ape species dispersed back to Africa. This lineage eventually *speciated* into *gorillas, chimpanzees* and *humans*.

This *theory* marks a *significant departure* from the long-held *view* that the evolutionary history of the lineage leading to the humans was *confined* to the African contenent. A theory similar to Stewert and Disotell's was *proposed* more than **25 years ago** by pioneering molecular anthropologist Vincent Sarich, but he *lacked* the *rigorous* analytic methodology necessary to prove it.

Stewert and Disotell's research is based on *parsimony* analysis. That is, the model which involves the *fewest* evolutionary events to explain the *data* is the most *plausible*.

The technique – which was *first* developed by anthropologists – uses computer technology to analyze *large* sets of data and identify the *most* parsimonious evolutionary model.

The Naked Ape: Human Behavior In A Zoologist's Eyes *Human behavior* is *animal behavior*, says *Dr. Desmond Morris*. Every behavior – *anger, attack, protection* – can be *found* in the jungle. His *groundbreaking book*, ***The Naked Ape,*** the *first* in a *trilogy* on human behavior, has just been *translated* into Chinese, and its *insights* into the *bloodlines* between *man* and *animal* are as *relevant* today as they were when it was published in **1967**.

"There are **193** species of monkeys and apes, **192** of them are *covered* with **hair**. The *sole* exception is *a naked ape* that calls itself **Homo sapiens**," said Desmond Morris in his controversial **1967** *best seller*, ***The Naked Ape***.

That *radical* premise *unleashed* a *furor* of a *similar magnitude* in the one *Charles Darwin* created when he suggested that *man had evolved from* the *Ape*, turning the book into a **12-million** copy *bestseller* and making a *star* of the *mild-mannerd* zoologist who authored it. Now, about **36 years** after it *shook up* readers in the West, the *groundbreaking* classic has been *translated* into Chinese (*Wenhl Publishing House*, three books in a series, 16 years each).

On a *recent* visit to Shanghai, Morris, now **75**, discussed his life's work and his *favorite* subject – observing "the human animal."

The idea that all men are *essentially* animals is *topical* in the current *political climate*, says Morris, noting that "one of the problems we have today is that people have started to think that *cultures* are *different*." Although the book *coined* a new term for the English language and was *translated* into **23** languages, Morris, an *Oxford University* zoologist, recalls that it was initially *viewed* as a *bad* joke in *appalling* taste.

"I had published *studies* of a *wide variety* of *other* creatures which had been read by a handful of specialists and caused *little* or *no* controversy," says the *affable* scientist, writer and artist. "But when I turned to writing books of *bare-skinned primates*, everything *changed*. I had assumed that *most* people were *ready* to face the fact that we are an *intergral* part of *primate evolution*. But I found myself *fighting* a *reargument* action for Charles Darwin. In some parts of the world *The Naked Ape* was *banned* and *illicit* copies were *confiscated* and *burned* by the *Church*. But things have *softened* a little bit now."

The *highly* acclaimed BBC television series "The Human Animal" has been a testimony to the *enduring* appeal of Morris' *original* view that human beings can justifiably be regarded and studied as just *one* of *many* animal species.

"*The Naked Ape* was written with the *belief* that precisely by *applying* these methods to human behavior, we could *gain* some interesting *insights* into ourselves," Morris asserts.

Inspired by this *new* mythology, Morris followed The *Naked Ape* with *Human Zoo*, which *focuses* on *human behavior* in *cities* – which says Morris, should not be called "concrete jungles." In their *natural habitats*," he explains, "wild animals do not *mutilate themselves, masterbate, attack their offspring, develop stomach ulcers*, or *commit murder*. Among the inhabitants of the concrete jungle, however, *all* of these behaviors *occur*. Man is *trapped* by his own *brainy brilliance* in a huge *menagerie* where he is in *perpetual* danger of *cracking* under the strain."

"Human beings have *evolved* over over a **million years** now, first living in groups of *only* **80** to **100** people – that's the *natural size* of a human community. But here in *Shanghai*, you have **16 million** people living *together*," he says. "*Human Zoo* concerns itself with how we *survive* in the *big city*."

Morris' *final* book in the *trilogy*, which addresses *personal* relationships in the city, is *Intimate Behavior*.

Considering *humans* as *animals* and *observing* them as *such* came to Morris when he was the *victim* of his *own* human *faults*.

"I was **38,** and really *pushing* myself at the time, almost doing six jobs: radio, TV, writing books, curatorship, lecture and research," he recalls. Basically, I worked *too* hard and *finally cracked*. That was the *wake-up call*: I realized that I was not a *machine*. I simply couldn't *push* myself so hard. I started to think really hard about my limitations, which *led* to my thinking of *human beings* as *animals*. It was then I started writing *The Naked Ape*, and finished it in four weeks."

And if *that* writing came easily – Morris is an unusually engaging, fluent writer – it is because *writing* is in his *blood*. His father, *Harry Morris*, was an *author* of children's fiction, and his grandfather, *William Morris*, founded Britain's *first penny paper*, the *Swindon Advertiser* (now the *Evening Advertiser*) in **1854**.

His *great-grandfather* was an an *important influence* on Desmond's life. The family moved to Swindon when Desmond was **5 years old**, and he *inherited* from his great grandfather *not* only a *microscopic* and strange scientific *specimens* he found in the family attic, but also the *curiosity* of a *naturalist.*

As a schoolboy, Desmond spent days *observing* the wildlife *in* and *around* the lake at *Queen's Park*, which was then *owned* by his grandmother. He became a student of zoology at *Birmingham University*, and then was *offered* a *research post* at *Oxford.*

As one of the world's *leading* authorities on animal behavior, Morris has appeared in **500** episodes of animal programs like "Zoology" for Granada and **100** episodes of "Life in the Animal World" programs for BBC.

But the man who *made* his name writing about naked apes confesses that he thinks of himself as neither *zoologist* nor *writer*: "I think of myself *as* a (surrealist) *painter*," he confesses. "I am *devoted* to it. But nobody *buys* my paintings simply because I *refuse* to *change* my *painting style*. So I make a living by studying zoology," says Desmond, one of the world's most famous zoologists, *almost* apologetically.

Morris *is self-deprocating* about his art, but he is nevertheless recognized as a *noted* m*odern artist*: In **1950**, he exhibited *jointly* with renowned Spanish surrealist painter *Joan Miro*, and has *exhibited* around the world.

Morris may not have been willing to *risk* becoming a *full-time* artist, but even his work in human behavior requires an *appetite* for risk: He had been *threatened* by the m*afia* when *filming* a body language *feature* on the streets of Italy and "nearly got into very serious *travel*" when filming *witch-doctors* in African countries.

But "does it, "he says, "because *observing* human behavior gives me a great deal of i*nsight* into human *problems* – and *ultimately*, their *solutions*."

At **75**, Morris shows *no* signs of *slowing* down. He spends most of his *conversations* r*elating* anadotes from his travels around the world – China is the **93**rd country he's been to.

In fact, *travel* is the latest field to *benefit* from his behavioral analysis. His **53**rd book is the *Naked Eye. Travels in Search of the Human Species*, and is *filled* with *monologues* and *autobiographical* and *natural* essays.

BIBLIOGRAPHY

Carrol, S. (2005), *Endless Forms Most Beautiful.*New York: W.W. Norton. ISBN 0-393-06016-0.
Charlesworth, C.B. and Charlesworth, D. (2003). *Evolution.* Oxford University Press. ISBN 0-192-80251-8.

Dawkins, R. (2006). *The Selfish Gene: 30th Anniversary Edition. Osford University Press.* ISBN 0199291152-30700-X.

Jones, S. (2001). *Amost Like a Whale: The Origin of Species Updated* (American Titie: *Darwin's Ghost).* New York: Ballantine Books. Asbn 0-345-42277-5.

Maynard Smith, J. (1993). *The Theory of Evolution: Canto Edition.* Cambridge University Press. ISBN 0-521-451-28-0.

Smith, C.B. and Sullivan, C.(2007). *The Top 10 Myths about Evolution.* Promotion Books. ISBN 978—1-59102-479-8.

Larson, E.J. (2004). *Evolution: The Remakable History of a Scientific Theory.* New York Modern Library, ISBN 0-679-64288-9.

Zimmer, C. (2001). *Evolution: The Triumpth of an Idea.* London: Harpar Collins. ISBN 0-060-19906-7.

Barton, N.H., Briggs, D.E.G., Eisen, J.A. Goldstein, D.B. and Patel, N.H. (2007). *Evolution,* Cold Spring Harbor Laboratory Press. ISBN 0-879-69684-2.

Cayne, J.A. and Orr, H.A. (2004). *Speciation.* Sunderland: Sinauer Associates. ISBN 0-878-93089-2.

Futuyma, D.J. (2005). *Evolution.* Sunderland: Sinauer Associates. ISBN 0-878-93187-2.

Gould, S.J. 2007). *The Structure of Evolutionary Theory.* Cambridge: Belknp Press (Harvard University Press). ISBN 0-674-00613-5.

Maynard Smith, J.. and Szathmary, E. (1997). *The Major Transition in Evolution.* Oxfordshire: Oxford University Press. ISBN 0-198-50294-X.

Mayr, E.(2001). *What Evolution Is.* New York: Basic Books. ISBN 0-465-04426-3

Futuyma, Douglas J. (2005). *Evolution.* Sunderland, Massachusetts: Sinauer Associates, Inc. ISBN 0-87893-187-2.

Lande R, Arnold SJ (1983). "The measurement of selection on corrected characters". *Evolution.* 37: 1210-26. doi:10.2307/2408842.

Ayala EJ (2007). "Darwin's greatest discovery: Design without designer". *Proc. Natl. Acad. Sci. U.S.A.* **104 suppl 1**: 8567-73.

Ian C. Johnson (1999). History of Science: Early Modern Geology. Malaspina University-College. Retrieved on 2008-01-15.

Bowler, Peter J. (2000). *Evolution: The History of an Idea.* University of California Press. ISBN 0-674-00613-5.

Darwin, Charles (1859). *On the origin of Species*, 1st, John Murray, p. 1. Relocated earlier ideas were acknowledged in Darwin, Charles (1861). *On the Origin of Species*, 3rd, John Murray, p. xiii.

AAAS Council (December 26, 1922).AAAS Resolution Present Scientific Status of the Theory of Evolution. American Association for the Advancement of Science.

IAP Statement on the Teaching of Evolution. The Interacademy Panel on International Issues (2006). Retrieved on 2007-04-25. Joint Statement issued by the national science academies of 67 countries, including the United Kingdom's Royal Society.

Board of Directors, American Association for the Advancement of Science (2006-o2-16). Statement on the Teachingof Evolution. American Association for the Advancement of Science, from the world's largest general scientific society.

Kutcschera U, Niklas K (2004). "The modern theory of biological evolution: an expanded synthesis". *Naturwissenschaften* **91** (6):225-76. doi:10.1007/s00114-0515-y. PMID 15241603.

Storm RA, Frudakis TN (2004). "Eye colour portals into pigmentation genes and ancestry". *TrendsGenes* **28** (8): 327-32.

Pearson H (2006). Genetics: what is a gene?" *Nature* **441** (7092): 398-401. doi:10.1038/44398a. PMID 15262401.

Peaston AE, Whitelaw E. (2006). "Epigenetics and phenotypic variations in mammals". *Mamm. Genome* **17** (5): 365-74.

Pearson, H (2006), Genetics, what is a gene?", *Nature* **441** (7092): 398-401.doi:10.

38/441398a. PMID 16724031.

Elizabeth Penniai (2007). DNA Study Forces of What It Meant to Be a Gene", *Science*

316 (5831):1536-1557.doi:10.1126/science.316.5831.1556. PMID 17569836.

See eg Martin's Novak's *Evolutionary Dynamics.*

Gerstein MH, Bruce C. Rozowsky JS, Zheng D. Du J. Korbel JO, Emanuelsson D. Zhang

ZD, Weissman S, Snyder M (2007). "What is a gene, post-ENCODE? History and updated definition". *Genome Research* **17** (6):669-681. doi:10.1101/gr.6339607.PMID17567988.

Cavalier-Smith T.(1985). Eukaryotic gene nembers, non-coding DNA, and genome size. In Cavalier-Smith T., ed. The *Evolution of Genome Size* Chichester: John Wiley.

International Human Genome Sequencing Consortium (2004). "Finishing the euchromatic sequence of the humangenome."*Nature* **431** (7011): 931-45, doi:1038/nature03001. PMID 15496913.

Penniai, Elizabeth (2007). "Working the Gene (Gene Count) Numbers. Finally, a firm Answer". *Science* **316** (5828):1113.

Watson JD, Baker TA, Bell SP, Gann, A, Levine M, Losick R (2004). *Molecular Biology of the Gene*, 5th ed. Peason Benjamin Cummings (Cold Spring Harbor Laboratory Press). ISBN 080534635X.

Vries, H. D (1889) Intercellular pongensis [1] ("pangen"definition on page 7 and 40 of this 1910 translation in English).

Mark B. Gerstein *et al*, "What is a Gene, post-ENCODE? History and updated definition," *Genome Research* 17(6) (2007):669—681.

Min Jou W, Haegeman G, Ysebacrt M, Fiers W (1972). "Nucleotide sequence of the genepool coding for the bacteriophage M52 coat protein". *Nature* **237** (5350): 82-8.1038/237082a0. PMID 4555447.

The Human Genome Project Timeline. Retrieved on 2006-09-13.

Darwin C. (1868). Animals and Plants under Domestication (1868).

Rassoulzadegan M, Grandjean V, Gounon P, Vincent S, Gillot l, Cuzin F (2006). "RNA-mediated non-mendelian inheritance of an epigenetic change in the mouse". *Nature* **441** (7092): 469-74. doi:10.1038/nature04674059.

Mortazavi A, Williams BA, McCue K, Schaeffer L, Wold B (May 2008). "Mapping and qualifying mammalian transcriptomes by RNA-Seq". *Nat. Methods*. Doi:10.1038/nmeth.1226. PMID 18516045.

Woodson SA (1998). "Ironing out the kinks: Splicing and translation in bacteria". *Genes Dev.* **12** (9): 1243-7.

Braig M. Schmitt C ((2006). "Oncegene-induced senescence: putting the brales on tumor development". *Cancer Res* **66**(9): 2881-4. doi:10.1158/0008-5472-CAN-5-4006. apmid 16540631.

Mount, DW (2004). *Bioinformatics: Sequence and genome analysis,* 2nd ed., Cold Spring Harbor Laboratory Press: Cold Spring Harbor, New York. ISBN 0879697121.

Lodish H, Berk A, Matsudaira P, Kaisar CA, Krieger M, Scott MP, Zipursky SL., Darnell J. (2004). *Molecular Cell Biology*, 5th, New York:WH Freeman.

Thomas KR, Capecchi R. Site-directed mutagenesis by gene targeting in mouse embryo-derived stem cells. Cell 1987;51;503-12 The 2007 Nobel Prize in Physiology or Medicine – Press Release

Deng C. In Celebration of Dr. Mario R. Capscchi's Nobel Prize. Int J Biol Sci 2007;3:417-419.

Lolle & colleagues (2005) Genome-wide non-mendelian inheritance of extra-genomic information in Arabidopsis. PMID 15785770.

Spilianakis & colleagues (2005) Interchromosomal associations between alternatively expressed loci. PMID 15880101.

Parra & Colleagues (2006) Tandem chimerism as a means to increase protein complexity in the human genome. MID 116344564.

Kapranov & Colleagues (2005) Examples of the complex architecture of the human transscriptome revealed by RACE and high-density tiling arrays. PMID 15998911.

Mark Collard, Bernard Wood: *How reliable are human phylogentichypothesis about human evolution are unlikely to be reliable* http//.pnas.org/content/9/500.abstract.

Leakey, Richard (1994). *The Origin of Humankind,* Science Masters Series, New York, NY:Basic Books, pages 87-89. ISBN 04650531.

Human Ancestors Hall: Homo neanderthalensis". Retrieved on 2008-05-26

Darwin, Charles (1861), *On the Orgin of Species,* 3rd, John Murray, **488**.

Dart RA (1925). "The Man-Ape of South Africa". *Nature* **115**: 195-199.doi:101038/11595a0.

Wood B (1996). "Human Evolution". *Bioessys* **18** (12):945-54. doi:10.1002/bies.950181204. PMID 8976151.

Wood B (1992). "Origin and Evolution of the genus Homo". *Nature* **355** (6363): 783-90. doi:10.1038/355783a0. PMID 1538759.

Cola-Conde CJ, Ayala FJ (2003). "Genera of the human lineage". *Proc. Natl Acad Sci. U.S.A.* **100**(13): 7684-9. doi:10.1073/pnas.0812372100. PMID 13244650.

HALDANE: JB (July, 1955). "Origin of Man". *Nature* **176** (4473): 169-70. PMID 13244650.

Kordos L, Begun DR (2001). "Primates From Rudabanya: allocation of specimens to individuals, sex and age categories". *Hum. Evol.***40**(1): 17-39. doi:101006/jhev.2000.0417. PMID 11139358.

Chimps are human, gene study implies – 19 May 2003 – New Scientist.

McBreaty S. Jablonsky N.G. de Lundey M. (2005). "First Fossil Chimpanzees". *Nature* **437**: 105-108. doi:10.1038/*nature*04008.

Strait DS Grine FE, Moniz MA (1997). "A Reappraisal of early hominid phylogeny". *J. Hum. Evol.* **32**(1): 17-82. doi:10.1006/jhev. 19960097. PMID 9034954.

Richard Leakey, Roger Lewin (1997). *Origins Reconsidered.* Little, Brown & Co. ISBN 0 349 10345 3.

Nimitz, C (2002). "A Theory on the Evolution of the Habitual Orthograde Human Bipedalism – The "Amphibisce Generalistheorie". *AnthropologischerAnzeiger* **60**: 3-66.

Crawford, M et al (2000). "Evidence for the unique function of docosahexanoic acid (DHA) during the evolution of the modern homonid brain". *Lipids* **34**: S-39-S47

Ungar, Peter S. (2006). *Evidence of the Human Diet: The Known, the Unknown, and The Unknowable.* US: Oxford University Press, 432. ISBN 0195183460.

Spoor F, Wood B, Zonneveld F (1994). "Implications of Early hominid labyrinth morphology for evolution of human bipedal locomotion". *Nature* **369** (6482). 645-8. doi:10.1038/369645a0.PMID 8208290

Carbonell, Eudald; Jose M Bermudez de Castro *et al* (2008-03). "The first hominin of Europe". *Nature* **452**: 465-469. doi:10.1038/nature06815. Retrieved on 2008-03-26.

Krigs M, Stone A, Schmitz RW, Kraintzki H, Stoneking M, Paabo S (1997). "Neandertal DNA sequences and the origin of modern humans". *Cell* **90**(1): 19-30. doi:10.1016/S0092-8674(00)80310-4. PMID 9230299.

Serre D, Langancy A, Chech M, *et al* (2004). "No evidence of Neandertal mtDNA contribution to early modern humans". *PloS Biol* **2** (3): E57. doi:10.1371/journal.pbio.0020057. PMID 15024415.

Gibbons, Ann (1998). "Solving the Brain's Energy Crises". *Science* **280** (5368): 1345-47. doi:10.11/science.280.5368.1345. PMID 9834409.

Ambrose SH (2001). "Paleplithic technology and human evolution". *Science* **291** (5509): 1748-53.PMID 111249821.

Mcbrearty S. Brooks AS (2000). "The Evolution that wasn't: a new interpretation of the origin of modern human behavior". *J. Hum. Evol.* **39** (5): 453-563. doi:10.1006/jhev.2000.0435. PMID 11102266.

Christopher Stringer and Peter Andrews (1988) "Genetic and Fossil Evidence for the origin of Modern Human" in *science* 239-1263-1268.

Rebecca L. Cann, Mark Stoneking, Allen C. Wilson (1987) "Mitochondrial DNA and human evolution" in *nature* 325: 31-36).

The Complete Works of Darwin Online – Biography. *Darwin-online.org.uk.* Retrieved on 2006-12-15.

The Mount House, Shrewsbury, England (Charles Darwin), Baruch College – Darwin and Darwinism. Retrieved on 2006-12-15.

"Darwin Correspondence Project-Letter 719 Darwin, C.R. to Fox, W.D. [15 June 1838]". Retrieved on 2008-02-08.

"Darwin Correspondence Project-Letter 729 – Darwin, C.R. to Hooker, j.D., [11 January 1844]" Retrieved on 2008-02-08.

"Darwin Correspondence Project-Letter 734 – Hooker, J.R., 29 January1844". 1844". Retrieved on 2008-02-08.

"Darwin and Design: historical essay". Darwin Correspondence Project (2007). Retrieved on 2008-09-17.

Darwin Correspondence Project: Inroduction to the Correspondence of Charles Darwin Volume 14. *Cambridge University Press.*

See list of books at Nineteenth Books on Evolution and Creation: scientific and religious debates in the age of Darwin.

The Children of Charles & Emma Darwin. *AboutDarwin.com.* Retrieved on 2006-12-15.

O'Connor, John J. & Robertson, Edmund F., "Charles Darwin". *MacTutor History of Mathematics archive.*

Edwards, A.W. F. 2004. Darwin, Leonard (1850-1943). In: *Oford Dictionary of National Biography, Oxford University Press.*

"Territory origins". Northern Territory Department of Planning and Infrastructure, Australia. Retrieved on 2006-12-15.

Charles Darwin National Park. Northern Territory, Australia Government. Retrieved on 2006-12-15.

Darwin College: About Darwin. Darwin College, Cambridge University website. Retrieved on 2006-12-15.

Darwin: A Life's Work. American Museum of Natural History. Retrieved on 2006-12-01.

Good Religion needs good science Rev Dr Malcolm Brown, Director of Mission and Public Affairs, Church of England. Retrieved 17 September 2008.

www.ingramcontent.com/pod-product-compliance
Ingram Content Group UK Ltd.
Pitfield, Milton Keynes, MK11 3LW, UK
UKHW060119300726
14090UKWH00002B/263

* 9 7 8 1 4 5 0 0 1 2 2 0 1 *